Electrochemical Engineering of Landfill Leachate

By
Valentino

Contents

Chapter 1

Introduction

1.1 General

Municipal Solid Waste (MSW) management is a global environmental challenge due to the rapid urbanization, population growth, and unsustainable consumption patterns. The global generation of MSW has been steadily increasing, with estimates suggesting that it will rise from 2.7 billion tonnes in 2020 to nearly 4.5 billion tonnes by 2050 [1]. As per the Central Pollution Control Board (CPCB) 2021–22 report the per day solid waste generation in India is 1,70,339 TPD and the annual generation is 62.2 million tonnes. The annual generation of MSW in India is projected to exceed 165 million tonnes by 2030, reflecting the growing stress on urban waste management infrastructure [2].

Among the various waste disposal strategies, landfilling remains the most widely practiced method, particularly in developing countries, due to its economic feasibility and ease of implementation [3]. However, a major by-product of landfill is the generation of landfill leachate a highly complex and contaminated effluent resulting from the percolation of water through waste layers. The leachate contains a diverse mix of organic and inorganic contaminants, including ammoniacal nitrogen (NH_4–N), total nitrogen (TN), heavy metals, chlorides, recalcitrant organics etc. Its composition varies with landfill age, waste composition, climate, and operational practices [4].

With time, the leachate undergoes various degradation phases namely, aerobic, acetogenic, methanogenic, and stabilization phases. Stabilized leachate, typically associated with landfills older than 10 years, and is characterized by a low BOD/COD ratio (< 0.1), indicating the presence of biologically non-degradable compounds and refractory organics like humic and fulvic like substances. In addition to the organic load, ammoniacal nitrogen is of great environmental concern in landfill leachates. It is released from the wastes, mainly by decomposition of proteins, and it has been found in leachates. Ammoniacal nitrogen concentration does not have an obvious decreasing trend with time, except due to dilution effects, and may disrupt biological units for leachate treatment due to its toxicity. This renders conventional biological treatments ineffective, necessitating the use of advanced physico-chemical and electrochemical treatment technologies for leachate treatment. In India, although there are no official statistic reports on the exact share of stabilized leachate, the disposal patterns and remediation progress suggest that most sites are legacy or aged, and only 17–20% of legacy waste had been remediated by mid of 2024. Combined with historical dependence on open dumping (77% of MSW) data, it can be concluded that well over 60 to 70% of the Indian leachate is mature or stabilized. Internationally, a meta-analysis of tropical countries (145 sites) found 75% of studied sites were old, consistent with the global trend that stabilized leachate predominates where landfills exceed a decade of operation [5].

Beyond direct risks from surface discharge, leachate percolation poses a critical threat to groundwater resources. Case studies across India demonstrated an elevated pollutant concentration in groundwater near landfills. For instance, at the Gazipur landfill in Delhi, COD in groundwater ranged 2–17 mg/L, with total dissolved solids (TDS) up to 2,208 mg/L, and iron levels exceeding permissible limits in several wells [6]. In three North Indian cities Chandigarh landfill, Panchkula and Mohali landfills groundwater near dumpsites showed COD in groundwater ranged from not detected to 128 mg/L (pre-monsoon) and to 80 mg/L (post-monsoon). Ammoniacal-N was 0.4–9.8 mg/L (pre-monsoon) and 0.09–6.8 mg/L (post-monsoon), Chloride: 15–115 mg/L (pre-monsoon), 9–67 mg/L (post-monsoon). Sodium: 39–74 mg/L (pre-monsoon), 19–53 mg/L (post-monsoon). Potassium: 17–40 mg/L (pre-monsoon), 12–43 mg/L (post-monsoon) [7]. Similarly, seasonal studies around dumpsites in western India reported groundwater deterioration in both pre- and post-monsoon conditions, with heavy metals and salts exceeding drinking water standards [8]. Such findings emphasize that landfill leachate is not only an on-site treatment challenge but also a persistent source of subsurface contamination with direct implications for human health and environmental safety.

1.2 Motivation for the Present Study

As per the CPCB Annual Report (2021–22) [9], urban local bodies reported 1,70,339 TPD of MSW generated and 1,56,449 TPD collected nationwide, reflecting progress in collection efficiency but leaving a sizeable quantity of waste uncollected. From the total waste generated, 91,511 TPD (54%) is processed or treated, 41,455 TPD (24%) is disposed of through sanitary landfills, and the gap in solid waste management is found to be 37,373 TPD (22%), underscoring the systemic leakage of waste from collection to scientific end-of-life pathways. The CPCB assessment highlights state-wise gaps in environmental monitoring at processing and disposal facilities and explicitly lists whether leachate quality is being monitored by State Pollution Control Boards (SPCBs) or Union Territory (UT) Pollution Control Committees (PCCs) at city facilities. Several jurisdictions reported "No" for leachate monitoring even where waste is landfilled [9]. Only about one in sixteen listed municipalities or facilities report conducting leachate quality monitoring, even though many landfills, dumps, or processing sites can generate leachate. In CPCB's Annual Report 2021–22, the two Kerala facilities marked 'Yes' for leachate quality monitoring are Composting Plant, Kochi and Composting Plant, Brahmapuram. Thus, for the year 2021–22, the monitored leachate entries reported for Kerala pertain to composting facilities rather than sanitary landfills. Systematic reviews of populations living near landfills have reported elevated risks of certain adverse health outcomes (e.g., some birth defects, selected cancers, and low birth weight), while also noting study limitations. The evidence nevertheless supports a precautionary approach to the control and monitoring of leachate [10-12].

Mature landfill leachate routinely operates outside the stable window of stand-alone biological units and shows site/season variability that destabilizes single-step treatments [13-16]. Pushing conventional physico-chemical options to regulatory endpoints often incurs sludge and energy penalties or rapid fouling. These constraints motivate oxidation-centric treatment that electrogenerated oxidants in situ and are adaptable to influent variability. Emerging literature also highlights the presence of Polyfluoroalkyl Substances (PFAS) and other persistent toxicants in landfill leachate globally, reinforcing the need for advanced treatment methods and routine surveillance for difficult-to-remove contaminants.

Conventional physico-chemical processes, such as coagulation-flocculation, adsorption, and membrane filtration, have been explored for stabilized leachate. While they can achieve partial pollutant removal, they face challenges such as sludge production, incomplete degradation of organics, high operating costs, and membrane fouling [11-12]. Given these limitations, researchers have increasingly focused on Advanced Oxidation Processes (AOPs), particularly Electrochemical Advanced Oxidation Processes (EAOPs), which can generate powerful oxidizing agents such as hydroxyl radicals (•OH) in situ [17-19]. These processes enable the breakdown of refractory organics into simpler, more biodegradable compounds [20] and simultaneously facilitate ammonia removal via indirect oxidation pathways [21].

The sustainability and cost-effectiveness of EAOPs are influenced by the materials used, particularly the electrodes. Traditional graphite-based cathodes or carbon-based cathodes, although effective, are often derived from non-renewable resources and involve energy-intensive fabrication processes [22]. There exists a pressing need to explore renewable, low-cost, and high-performance electrode materials to enhance the sustainability of electrochemical treatment systems.

This study is motivated by (i), India's documented 22% MSW handling gap in 2021–22 and the concomitant risks from inadequately monitored and treated leachate (ii), the intrinsic recalcitrance of stabilized leachate that limits biological routes and (iii), the sustainability gap in conventional cathode materials. In response, evaluate a sequential Aerated Electrocoagulation (AEC) -Peroxicoagulation (PC) method for stabilized landfill leachate and develop a renewable biomass-derived carbon (BDC) cathode (from waste teakwood) to enhance in-situ H_2O_2 generation. The integrated process is assessed for pollutant removal (COD/TOC/NH_4^+-N), sludge yields (TS/VS), energy/charge metrics, and by-product risks, with the aim of delivering an oxidation-centric, scalable, and materially sustainable pathway aligned with Indian municipal needs [9-22].

1.3 Research Rationale

The treatment of stabilized landfill leachate has become a persistent challenge due to its contaminant quality and the quantity. Over the years, several conventional and advanced processes have been investigated, but when applied individually they exhibit notable shortcomings. Biological treatment methods and physico-chemical treatment, although cost-effective and widely adopted for leachate treatment, they show poor performance in stabilized leachates [14,17,23,24]. Similarly, electrochemical advanced oxidation processes, when applied as standalone treatment, are restricted by electrode passivation, limited current efficiency, and high energy consumption, which hinder their practical scalability [25,26]. These limitations are tied to the way iron phases evolve under operating conditions and to how dissolved oxygen (DO) and electrolyte chemistry govern the formation of mixed-valent iron hydroxides during treatment [27,28]. These limitations underline the importance of hybrid and sequential processes that can combine complementary mechanisms to achieve higher treatment efficiency and long-term stability. Among EAOPs, AEC and PC have gained prominence in recent years. AEC enhances pollutant destabilization while minimizing electrode passivation [29] and PC introduces hydrogen peroxide and additional iron ions, facilitating advanced oxidation [30].

In this context, aerated electrocoagulation has emerged as a practical refinement of conventional iron electrocoagulation (EC). In the standard process, sacrificial iron dissolves to form iron hydroxide flocs that destabilize colloidal and suspended matter [31,32]. A persistent operating issue is passivation of the iron surface by oxide / magnetite films that curtail metal dissolution and active surface area [25]. Introducing aeration during EC improves hydrodynamics and shifts a portion of the ferrous to ferric oxidation into the bulk solution, which mitigates passivation and stabilizes performance; recent work on aerated iron EC reports consistently higher removals than the non-aerated mode and, under suitable conditions, the appearance of additional oxidative pathways including ferryl (Fe(IV)) species [29]. These observations align with mechanistic studies of Fe(0) electrolysis showing that dissolved oxygen, pH and electrolyte anions govern whether green rust or magnetite predominates, with narrow O_2 windows separating Fe(II, III) from Fe(III) hydroxide formation and with chloride favouring magnetite while bicarbonate favours carbonate green rust [28]. Complementary phase-resolved work confirms that mixed-valent nanoparticles produced by Fe(0) electrocoagulation evolve between magnetite and green-rust domains as current density and electrolyte conditions change, linking operating conditions to both floc chemistry and passivation behaviour [27]. In applied reactors, simple gas sparging has been shown to enhance EC performance.

Aeration/gas sparging improves mass transfer and dissolved-oxygen availability and can mitigate (but not eliminate) electrode passivation [33], which explains the operational advantage observed for AEC.

The cathode choice is equally important because, in EAOPs that rely on the oxygen reduction reaction, the cathode governs in-situ H_2O_2 generation and thus radical production [26,34]. An effective cathode combines high electrical conductivity with a large, accessible electroactive surface, demonstrates selectivity for the two-electron oxygen reduction pathway, and remains chemically and electrochemically stable under oxidizing conditions while retaining mechanical integrity over extended operation [34]. Practical deployment favours robust, non-corrosive and affordable carbons; recent reviews and applications show that biomass-derived carbon electrodes provide hierarchical porosity, heteroatom-doped sites and adequate conductivity that together enhance adsorption and electrocatalysis, offering a sustainable alternative to conventional graphite in peroxicoagulation / electro-Fenton (EF) configurations [35-37]. Therefore, the sequential integration of AEC and PC has emerged as a promising approach, offering synergistic benefits by combining pollutant destabilization with oxidative degradation. This hybridization has the potential to achieve higher removal efficiencies, reduce sludge generation, and improve long-term treatment stability.

1.4 Research Objectives

In light of the aforementioned challenges and opportunities, the present thesis seeks to:

1. Evaluate the performance of standalone and sequential electrochemical treatment processes, specifically aerated electrocoagulation and peroxicoagulation, for the treatment of stabilized landfill leachate.
2. Develop and characterize a novel biomass-derived cathode electrode synthesized from waste wood using pyrolysis process, and to assess its electrochemical properties.
3. Evaluate the performance of BDC as cathode electrode in the electrochemical treatment of leachate and to compare its performance with graphite cathode.

1.5 Scope of the Study

This study examines the treatment of real, stabilized landfill leachate collected from a municipal solid waste dumpsite in Kerala, India. The experimental work encompasses the design and operation of a bench-scale electrochemical reactor and the synthesis and characterization of biomass-derived carbon cathodes produced from teak wood waste. The treatment is evaluated under three configurations: standalone aerated electrocoagulation, standalone peroxicoagulation process, and sequential operation in both orders (AEC-PC and PC-AEC). In the AEC stage, iron is used as the sacrificial anode with an iron counter-electrode as the cathode, and continuous aeration is provided to sustain coagulation. In the PC stage, an iron anode is paired with a carbon-based cathode (commercial graphite or the synthesized BDC).

Across all configurations, the influence of key operating variables such as pH, current density, electrode material, and reaction time is systematically assessed. Analytical endpoints include conventional pollution parameters such as COD, TOC, TN, and NH_4–N, while sludge metrics are also evaluated. Material characterization of the BDC electrode employs Scanning Electron Microscopy (SEM), Energy Dispersive X-ray (EDX), Raman spectroscopy, Fourier Transform Infrared Spectroscopy (FTIR), Brunauer-Emmett-Teller (BET) surface area analysis, and electrochemical impedance spectroscopy to link surface chemistry and morphology with process performance. The comparative analysis of standalone versus sequential operation and of graphite versus BDC cathodes is also conducted, defining the process window and providing evidence for a practical, scalable treatment method for stabilized leachate.

1.6 Organization of the Thesis

This thesis is systematically organized into six distinct chapters, detailed as follows:

Chapter 1 introduces the problem context, motivation, rationale, objectives, scope, and the overall structure of the thesis.

Chapter 2 presents a critical literature review covering MSW and leachate management; conventional biological and physicochemical treatments; AOPs and their limitations;

EAOP fundamentals with emphasis on AEC and PC; electrode materials; and the state of biomass-derived carbon electrodes, culminating in explicit research gaps.

Chapter 3 details the methodology; materials and leachate characterization; reactor configuration; optimization protocols for AEC and PC; sequential treatment experiments; BDC fabrication and multi-technique characterization; analytical methods; and treatment performance evaluation metrics.

Chapter 4 reports results for standalone and sequential AEC and PC treatment; effects of pH and current density in AEC and PC; comparative performance of AEC-PC vs PC-AEC and sludge quantification.

Chapter 5 presents the development and characterization of the biomass-derived cathode and its application in PC and in sequential AEC-PC treatment process, including comparisons with graphite.

Chapter 6 synthesizes the findings, states the main conclusions, and outlines scope for future work, including pathways for scale-up and integration with other treatment units.

Chapter 2

Literature Review

2.1 Introduction

This chapter reviews existing research and global practices related to municipal solid waste generation, management strategies, leachate generation, and the treatment of leachate using advanced oxidation processes, particularly electrochemical treatment methods. The discussion also covers various innovative cathode materials used for the electro-generation of hydrogen peroxide in electrochemical processes. This review establishes a foundation for identifying the research gaps and the scope addressed in the present study.

2.2 Municipal Solid Waste and Leachate Management

Municipal solid waste generation continues to increase globally with the rise in household income and the expansion of urban areas. The global MSW generation was projected to reach around 4.5 billion tonnes by 2050 [1]. Despite gradual improvements in waste collection and treatment capacities, a significant portion of the world's MSW is still not managed within controlled systems. Nearly one-third of global waste continues to be improperly handled through open dumping and/or open burning, particularly in low-income countries [38-40].

High-income countries manage a larger fraction of their waste safely and exhibit relatively consistent practices. Collection coverage is nearly universal, materials are diverted through recycling and composting, residual waste is treated via waste to energy incineration and only the final remainder is disposed of in engineered sanitary landfills equipped with environmental controls. In contrast, low and lower middle income countries face significant gaps in collection coverage and continue to rely heavily on uncontrolled dumpsites, while recycling, composting, and energy recovery occur at comparatively low levels [38,41]. These operational differences are accompanied by distinct per capita waste generation levels. Analyses based on multi-country datasets show that per capita MSW generation is considerably higher in high-income economies (often approaching or exceeding 1 kg per capita per day) and substantially lower in low-income economies (typically only a few tenths of a kg per capita per day), with urbanization and income serving as the principal driving factors [42,43].

In India, MSW generation has increased from 42 million tonnes in 2011 to over 62 million tonnes at present and is projected to reach approximately 165 million tonnes by 2030. Within the country, per capita waste generation varies significantly with city size and income levels, typically ranging from about 0.1 to 0.5 kg per capita per day, with higher values observed in larger and wealthier urban centers. Recent cross-city assessments report ranges as high as 0.19 to 0.99 kg per capita per day across major metropolitan areas [44]. The composition of MSW in India remains predominantly organic-rich; for instance, city-scale audits conducted in Bengaluru indicate that approximately 55–65% of the waste stream comprises biodegradable (wet) waste, with the remainder consisting of paper, plastics, inerts, and other materials [44,45]. As per the CPCB 2021–22 report, approximately 24% of the total solid waste generated in India is landfilled. A total of 341 landfills is currently in operation for solid waste dumping, 11 landfills have been capped after reaching full capacity, 305 new landfills are under construction, and 1,924 landfill sites have been identified. Leachate generation is a major problem associated with the dumping of wet waste in landfills, and its management remains a serious concern.

Landfill leachate management rests on two pillars: (i) minimizing leachate generation through landfill hydrologic control and containment, and (ii) applying treatment methods suited to leachate age and composition. Landfill systems are typically designed with daily and intermediate covers, stormwater diversion structures, composite liners equipped with Leachate Collection and Removal Systems (LCRS), and monitoring wells to reduce infiltration, maintain low hydraulic head on liners, and channel leachate for controlled treatment or recirculation, thereby mitigating risks at the source [46]. Recent global syntheses emphasize that climatic drivers (precipitation and temperature) and waste composition (notably high organics) strongly influence leachate strength and risk profile, particularly COD, ammonia, chloride, metals, and xenobiotics, requiring site-specific combinations of biological and physico-chemical processes for its treatment [47]. In parallel, comparisons of leachates from different waste streams show composition can shift markedly with inputs, which directly affects treatability and the sequencing of unit processes [48].

Young and intermediate leachate (higher BOD/COD) benefits aerobic biological treatment options (e.g., ASP or SBR or MBR) with nitrification-denitrification for nitrogen removal; mature or stabilized leachate (low BOD/COD, humic and fulvic dominance, high NH_4–N and salts) typically requires oxidative and / or separation polishing step to meet discharge norms [46,47,49]. The treatment processes adopted for these leachates vary with their characteristics and the treatment technologies employed by different researchers are discussed in the following section.

2.3 Conventional Leachate Treatment Technologies

The different biological and physico-chemical treatment technologies applied for the treatment of landfill leachate is discussed in this section.

2.3.1 General

Landfill leachate is a complex wastewater containing high loads of organic matter, refractory organics, nutrients, inorganic salts, heavy metals and trace emerging contaminants; its composition varies with landfill age, waste composition, climate and site operation [17,50]. In practice, leachate is often classified by its age into young (< 5 years), intermediate (5–10 years) and stabilized (> 10 years) streams. Biodegradability falls with time, with typical BOD_5/COD ratios shifting from 0.4–1.0 (young) to 0.2–0.3 (intermediate) and ≤ 0.1 (mature or stabilised); total nitrogen and humic-like substances become more prominent in stabilized leachate [23,50]. Ammoniacal nitrogen (NH_4–N) frequently occupies the hundreds to several thousands of mg/L band in mature streams; values of 2,000–4,000 mg/L are regularly reported at aged sites, posing inhibition risks for nitrification [23,51].

Conventional biological processes perform well for young leachates but show lower efficiency as the leachate stabilizes. Aerobic activated-sludge, SBR, and MBR configurations can deliver high COD and nitrogen removal when readily biodegradable organics dominate and ammonium levels are moderate; for example, full-scale and pilot reports of MBR-based processes show near-complete nitrification with overall COD polishing when coupled with downstream membranes, especially above 15 °C and with long SRTs [52]. In mature leachates, where BOD_5/COD drops to ≤ 0.1 and NH_4–N rises into the hundreds to thousands of mg/L, biological units alone exhibit diminished COD removal and sensitivity to free-ammonia inhibition, motivating pretreatment or hybrid methods [23,50,51].

Within coagulation–flocculation, studies underline its role as a pretreatment for stabilized leachate; ferric chloride or alum at acidic pH reduces color, turbidity, and a portion of dissolved COD while improving downstream membrane performance. Amokrane et al. [50] optimized $FeCl_3$ vs alum and pH for leachate pretreatment before RO, showing that coagulation significantly lowers fouling propensity but leaves much of the dissolved organics in solution. Tatsi et al. [54] examined raw and partially stabilized leachates; $FeCl_3$ at acidic pH outperformed alum for color and turbidity removal and yielded composition-dependent COD reduction, which was effective as pretreatment but insufficient alone for stabilized leachate. In a stabilized leachate, Hasar et al. [55] measured only 36% COD and 46% Suspended Solids (SS) removal in the coagulation-flocculation step of a stabilized leachate treatment process, necessitating subsequent MBR and RO stages to meet discharge standards. Reported limitations are consistent across studies; high chemical demand and pH adjustment requirements, sludge generation, and the intrinsically limited removal of refractory dissolved COD. Coagulation–flocculation is most valuable as a conditioning step rather than a standalone solution for stabilized leachate [53-55].

For adsorption, Granular Activated Carbon (GAC) is effective at removing inert COD and trace metals in mature leachates, but its performance depends strongly on the carbon's pore structure and it quickly encounters bed saturation. Mohammad-Pajooh et al. [56] compared several GACs for stabilized leachate treatment and showed substantial removal of inert COD and metals, while highlighting sensitivity to carbon properties and contact time. Kurniawan et al. [57] provided a useful benchmark on a very low-biodegradability leachate (BOD_5/COD = 0.08; initial COD 8,000 mg/L; NH_4–N 2,600 mg/L). GAC alone removed 58% COD, whereas integrated H_2O_2 oxidation + GAC achieved 82% COD and 59% NH_4–N removal under optimized conditions. Even then, the integrated approach failed to meet stringent COD (< 200 mg/L) and NH_4–N (< 5 mg/L) discharge limits, underscoring adsorption's limitations as a sole advanced step for mature leachate. Practical constraints include regeneration and disposal of spent carbon after adsorption, competition from humic-like NOM and limited direct removal of ammoniacal nitrogen.

Table 2.1 Limitations in Treating Stabilized Leachate

Treatment process	Limitation	Specific issue	Reference
Biological	Low biodegradability	Stabilized leachate shows BOD/COD < 0.1, limiting microbial degradation efficiency.	[52,58]
	Inhibition by leachate matrix	High ammonia and refractory organics inhibit microbial activity and biomass health.	[59,60]
	Nitrogen removal challenges	Inefficient NH_4–N removal, especially at elevated ammonia concentrations.	[61,62]
	Process sensitivity & time	Long treatment times; sensitivity to pH shifts and toxicants reduces stability.	[63,64]
Physicochemical	Cost intensity	High operational chemical costs for coagulants, adsorbents, and consumables.	[65]
	Membrane issues	Membrane fouling and brine generation necessitate frequent maintenance management.	[66,67]
	Incomplete polishing	Incomplete removal of micro-pollutants and humic substances; residual color and TOC may persist.	[68]
	Sludge handling	Sludge production in coagulation–flocculation requires dewatering and disposal.	[69]

For membrane processes, both pressure-driven systems (UF / NF / RO) and MBR-combined with UF/ NF/ ROs are widely deployed for stabilized leachates. The review by Ahmed & Lan [52] synthesizes a decade of MBR research and shows that MBRs can reliably remove BOD and most NH_4–N, producing solids-free effluents that reduce fouling in downstream NF and RO systems. When MBR effluent is polished by RO, overall COD removals of > 90–99% are routinely reported. In Hasar et al. [55] study, after UF, the aerobic or anoxic MBR achieved > 90% COD and > 92% inorganic-N removal (SRT 30 Days) and RO then reduced effluent COD to < 4 mg/L, clearly illustrating the need for sequential processes for mature leachate treatment. The persistent limitations of membrane treatments are also well documented, including fouling and cleaning cycles in MBR / NF / RO systems and especially the management of RO concentrate (brine) from leachate plants, which remains a major engineering and regulatory challenge [52].

Taken together, these studies show why stabilized leachate rarely yields to a single treatment and the limitations of treating stabilised leachate is given in Table 2.1. The coagulation-flocculation process is valuable for conditioning, adsorption can abate inert organics and metals, and membrane processes deliver the regulatory disposal norms after treatment, but each process has significant limitations in isolation. Consequently, the highest and most stable performance in mature leachate treatment comes from hybrid sequences that combine complementary mechanisms and target the specific mix of refractory organics and high ammoniacal nitrogen typical of stabilized landfill leachate.

This has shifted the focus towards advanced treatment methods including electrochemical processes and advanced oxidation processes which show promise in treating recalcitrant leachate but require further research to assess their scalability [3].

2.4 Advanced Oxidation Processes

Advanced Oxidation Processes represent a set of powerful oxidative techniques used for the treatment of a wide range of pollutants, particularly in leachate and wastewater treatment. AOPs rely on the generation of highly reactive hydroxyl radicals ($HO^•$) and sulfate ($SO_4^{•-}$) radicals, which have a high oxidation potential and are capable of breaking down a wide range of organic contaminants that are often resistant to conventional treatment methods. This section provides an overview of AOPs, their application in leachate and wastewater treatment, and the limitations of standalone AOPs. Hydroxyl radicals (E° 2.7–2.8 V vs NHE) are near-non-selective and react rapidly with most organics, whereas sulfate radicals (E° up to 3.1 V) are more selective electrophiles that react fastest with electron-rich substrates. The following are the different types of AOPs.

Advanced Oxidation Processes refer to a group of chemical treatment technologies that rely on the in-situ generation of strongly oxidative radical species primarily hydroxyl (HO•) and sulfate ($SO_4^{\bullet -}$) radicals for the degradation of recalcitrant organic contaminants in water and wastewater systems. Hydroxyl radicals (E° 2.7–2.8 V vs NHE) are near non-selective and react rapidly with most organics, whereas sulfate radicals (E° up to 3.1 V) are more selective electrophiles that react fastest with electron-rich substrates. The following are the different types of AOPs.

2.4.1 Types of Advanced Oxidation Process

The different AOP processes like Fenton and photo-Fenton, Ozone-based AOPs, Photocatalysis, Electrochemical AOPs, Sulfate Radical based AOPs etc., are discussed briefly in this section.

Fenton and photo-Fenton processes

Fenton chemistry uses Fe^{2+} to activate H_2O_2 and generate hydroxyl radicals ($HO^{\bullet}$), while photo-Fenton accelerates Fe^{3+}-Fe^{2+} cycling with light, so radical production is sustained; both routes work best in acidic conditions (pH 2.5–3) and are widely used as pre-biological conditioning for stabilized leachate because they abate COD and raise BOD_5/COD [70]. Recent real-leachate studies show this pattern clearly; a solar photo-Fenton-like treatment increased biodegradability (BOD_5/COD) and reduced toxicity irrespective of whether a coagulation pre-step was used, confirming its role upstream of biological polishing treatment (UVA, acidic pH, Fe-catalysis) [71] and a hazardous-waste leachate treated by photo-Fenton achieved 85.3% COD removal with 93.3% PAH removal under optimized iron and peroxide at acidic pH [72].

Ozone-based AOPs (O_3, O_3/H_2O_2, O_3/UV)

Ozone acts both as a molecular oxidant and in alkaline or peroxone conditions (O_3/H_2O_2) or with UV, as a precursor to $HO^{\bullet}$, in leachate treatment it was commonly placed upstream of bio-steps to open up recalcitrant matter and reduce color / UV-absorbing chromophores [73,74]. On real landfill leachate that had already undergone nitrification–denitrification and / or coagulation, O_3/UVC was the only variant that significantly enhanced biodegradability (> 70%) even though COD legal limits were not reached, an outcome that matches how operators use ozonation to make subsequent biological treatment feasible. Ozone-based oxidation on biologically treated leachate reports effective removal of recalcitrant organics with dosing guided by surrogate monitoring [75].

Photocatalysis (UV/TiO_2)

In photocatalysis, light-activated semiconductors generate electron-hole pairs and surface-bound oxidants ($HO^{\bullet}$) that degrade adsorbed organics. The performance depends on light delivery, turbidity of the wastewater, concentration of the pollutants, and catalyst contact (immobilized or recoverable) [76,77]. A recent real-leachate study using $CaTiO_3$ nanoparticles reported 72% COD removal with 43.62% color and 41.75% turbidity reduction at optimized dose and irradiation. The values representative of modern visible / UV systems on mature leachate, while the broader comparisons of doped visible-active TiO_2 on landfill leachate confirm the gains from optical tuning and reactor design when treating stabilized leachate [78].

Electrochemical Advanced Oxidation Process

Electrochemical AOPs, such as electro-Fenton and anodic oxidation, utilize electrical current to produce ($HO^{\bullet}$) at the electrode-solution interface. The electro-Fenton process is derived from the Fenton process, where Fenton's reagent, a combination of H_2O_2 and Fe^{2+} ions, is introduced externally to the wastewater undergoing treatment [79]. The Fenton method has been utilized for wastewater treatment since the 1980s; however, its advancement has faced challenges due to issues such as the requirement for significant chemical quantities, the production of process sludge, and reduced efficiency stemming from parasitic reactions [80,81]. To address the limitations of the Fenton process, one effective approach is to integrate it with electrochemistry, which provides an excellent method for the in-situ generation of H_2O_2 through the two-electron reduction of dissolved O_2. Alternatively, Fe^{2+} can be electrocatalytically regenerated from ferric iron, which is produced in the Fenton reaction Eqs. (2.1) and (2.2), thereby circumventing the need for large quantities of iron salts. The continuous generation of H_2O_2 and electro-regeneration of Fe^{2+} at an optimal operational pH condition mitigates expenses linked to chemical reagents, reduces the formation of process sludge, and diminishes the risks tied to the transport and storage of H_2O_2 [82]. Anodic oxidation entails the oxidation of pollutants through: (i) direct electron transfer to the anode surface (M), (ii) heterogeneous Reactive Oxygen Species (ROS) produced as intermediates of oxidation of water to oxygen, including the powerful physiosorbed $HO^{\bullet}$at the anode surface, denoted $M(HO^{\bullet})$, generated via Eq. (2.3), and weaker oxidants like H_2O_2 produced from $M(HO^{\bullet})$ dimerization by Eq. (2.4) and O_3 formed from water discharge at the anode surface by Eq. (2.5), and / or (iii) other weaker oxidant agents electrochemically produced from ions existing in the bulk solution [26]. In contrast to the anodic oxidation process (which is a direct electrooxidation process), the electro-Fenton is an indirect electrochemical process based on the generation of homogeneous $HO^{\bullet}$in the solution through Fenton reaction. A EF study on mature landfill leachate reports very high removals under optimized conditions, achieved 97.4% COD removal, colour 99%, NH_4–N 100%, and TN 94.2% with an energy use of 0.23 kWh/g COD [83]. For anodic routes, BDD electro-oxidation as pretreatment on raw leachate (influent COD 5,440 mg/L) operated at 40 mA/cm^2, pH 7.5, electrolysis time $\leq$ 250 min reached COD 93% removal and NH_4–N 82% removal and improved subsequent biological nitrogen removal [84].

$$Fe^{2+} + H_2O_2 \rightarrow Fe^{3+} + HO^{-} + HO^{\bullet} \quad (2.1)$$

$$Fe^{3+} + e^{-} \rightarrow Fe^{2+} \quad (2.2)$$

$$M + H_2O \rightarrow M(HO^{\bullet}) + H^{+} + e^{-} \quad (2.3)$$

$$2M(HO^{\bullet}) \rightarrow 2M + H_2O_2 \quad (2.4)$$

$$3H_2O \rightarrow O_3 + 6H^{+} + 6e^{-} \quad (2.5)$$

Sulfate Radical based Advanced Oxidation Process

In Sulphate Radical-based AOPs (SR-AOPs), peroxydisulfate (PDS) or peroxymonosulfate (PMS) are activated thermally, photochemically, electrochemically, or with mineral catalysts (e.g., pyrite) to form $SO_4^{\bullet -}$ (often alongside some $HO^{\bullet}$). Although $SO_4^{\bullet -}$ has a high standard potential of up to 3.1 V [85], it is more selective than $HO^{\bullet}$(faster with electron-rich aromatics, slower with saturated aliphatics) and has a wider workable pH range than the classical Fenton process [86]. On stabilized leachate, pyrite-activated PDS at an FeS_2 dosage of 1,000 mg/L, $Na_2S_2O_8$ dosage of 1,000 mg/L, and pH 8.8 delivered COD removal of 78.5% and TOC removal of 74.2%, with BOD_5/COD rising from 0.023 to 0.41. The quenching tests confirmed the contributions from both $SO_4^{\bullet -}$ and $HO^{\bullet}$ [87] on the treatment of organic matters. For the treatment of a biologically treated leachate, pyrite-activated PDS at pH 6, with a pyrite concentration of 9.28 mM and a dimensionless oxidant dose of 0.25, achieved 45% additional COD removal and produced an effluent compatible with autotrophic denitrification. These AOPs involve short-lived radicals that facilitate rapid oxidation and mineralization of otherwise non-biodegradable organic matter, making them especially suitable for complex wastewater matrices like landfill leachate [88].

2.4.2 Application of Advanced Oxidation Processes in Wastewater Treatment

Advanced oxidation processes have gained prominence in landfill leachate and industrial wastewater treatment due to their ability to degrade compounds that resist conventional biological or physicochemical approaches [89]. These include pharmaceuticals, PAHs, humic acids, and ammonia rich wastewater.

Fenton and photo-Fenton treatments are among the most studied AOPs for leachate treatment. Fenton and photo-Fenton process for stabilized leachate provided the stable results under the acidic range (2.5–3) of pH and when iron cycled efficiently. Under these conditions Fe^{2+} activates H_2O_2 to form $HO^{\bullet}$, and irradiation in photo-Fenton accelerates Fe^{3+} to Fe^{2+} regeneration and sustains radical flux. Laboratory and pilot-scale applications have consistently shown high removal efficiencies, with COD and color reductions exceeding 90% under optimal conditions [90]. A robust recent example on PAH-rich, hazardous-waste leachate reported 85.3% COD removal and 93.3% PAH removal under optimized iron and peroxide at acidic pH, with concurrent toxicity reduction [72] illustrative of what well-tuned Fenton or photo-Fenton can deliver on stabilised leachate. When the same reactions are examined at near-neutral pH the efficiency declines, because neutralization after reaction precipitates $Fe(OH)_3$, these routes unavoidably generate an iron-rich sludge that must be dewatered and managed. The contemporary work explicitly quantifies and seeks to minimize this burden, for instance by optimizing Fe/H_2O_2 to curb sludge yield and documenting sludge characteristics for safe handling. At the optimum molar ratio of Fe/H_2O_2 (1: 2) the COD, BOD_5, TN removals were 70%, 67% and 84%

respectively and the sludge to iron ratio and organic removal to sludge ratio were 1.147 L/mol and 33.79 g/L respectively [91].

Ozone systems are often applied as a pre-biological step to bleach colour and partially oxidize refractory organics while shifting dissolved organic matter towards more biodegradable fractions. On stabilized leachate, O_3 only typically delivers moderate COD abatement (30–50%) alongside marked colour removal (80–100%). For example, on biologically treated leachate Chys et al. [92] reported 87–100% colour removal at slightly alkaline pH using ozone alone, while in a classic O_3 only run Wang et al. [93] achieved 47% COD removal and 91% color removal at an ozone dose of 2.6 g/L, with BOD_5 increasing by 324% the kind of biodegradability lift operators target before a nitrifying bioreactor.

Photocatalysis with TiO_2 uses light to generate $HO^\bullet$and related oxidants, the performance of photocatalysis hinges on photon delivery through turbid matrices and on catalyst contact. On real leachate, TiO_2 systems have recently been shown to improve biodegradability and trim organic surrogates under practicable doses. A post-treatment study using Ag–TiO_2 reported maximum reductions of 62% COD, 66% color, 53% turbidity and 55% TOC after four hours of irradiation at its optimized conditions, reinforcing the role of TiO_2 as a conditioning step rather than a stand-alone finish on stabilized effluents [94].

The EF and anodic oxidation processes were increasingly used as pretreatments to raise biodegradability before biological steps, with operating windows managed in chloride-rich matrices to limit oxy-chlorine by-products. On membrane-concentrated landfill leachate, a recent EF study using Fe_3O_4-loaded GAC reported 95.7% COD removal and 94.9% TN removal at an energy consumption of 0.098 kWh/g COD, illustrating why EF was often placed upstream to “unlock” downstream biology process [95]. In a separate BDD anodic oxidation study on pretreated leachate nanofiltration concentrate (influent COD 4,900 mg/L; NH_4–N 21 mg/L), at optimum conditions (pH 5.5, 333 A/m^2, 8 h) achieved 86.4% COD removal and 94.4% NH_4–N removal, while also reporting by-product formation profiles that help set safe operating limits [96]. Sulfate-radical AOPs performs well on aromatic-rich, stabilized leachates and over a wider pH window than classical Fenton. On raw municipal landfill leachate from Morocco UV/Fe^{2+}/$S_2O_8^{2-}$

achieved up to 93% COD removal, 98% colour removal, and 77% UV_{254} reduction at optimized doses under acidic conditions, showing SR-AOPs as a pre-biological conditioning step to cut chromophores and apparent recalcitrance or as a chemical polish when discharge limits are tight [97].

Several recent studies emphasize the integration of AOPs with biological or membrane systems, which improves overall treatment performance and reduces operating costs. Sequential application, such as AOP followed by membrane filtration or biological degradation, has shown enhanced COD removal rates and reduced membrane fouling [24,26]. Across these routes, the consistent lesson was that AOPs earn their keep when integrated thoughtfully with other processes. Ozone, photo-Fenton and electro-Fenton prior to bio-oxidation can raise BOD_5/COD and suppress toxicity; electro-Fenton pretreatment can remove NH_4–N and oxidize humics to reduce downstream fouling; and membrane-coupled trains benefit from AOPs that reduces chromophores and low-humic acids like substance that otherwise poison RO/NF [98].

2.4.3 Limitations of Standalone Advanced Oxidation Processes

While AOPs are efficient in removing a wide array of recalcitrant compounds, their application as standalone treatments in full-scale systems is hindered by several limitations:

High operational costs: Many AOPs (e.g., ozonation, electrochemical oxidation, UV-Fenton) involve high energy and reagent consumption, particularly with ozone generators, UV lamps, or electrical systems. This directly affects process economics, especially at scale [99].

Sludge and secondary waste generation: Fenton-based systems produce iron sludge in considerable volumes, requiring dewatering and disposal. This adds to the cost and complexity of plant operation [90].

Scavenging effects: Radical scavengers such as bicarbonate (HCO_3^-) and natural organic matter present in landfill leachate can significantly reduce the efficacy of radical formation, reducing overall treatment efficiency [89].

Formation of toxic by-products: Partial oxidation may lead to the production of more toxic or persistent transformation products (e.g., aldehydes, ketones) which require additional treatment steps [100].

Scalability and process control issues: Most AOPs have demonstrated great efficiency in laboratory or pilot setups, but commercial-scale adoption remains limited. Factors such as energy efficiency, long-term reliability, and operator training influence scalability [99].

To address these limitations, current research promotes the development of hybrid AOP systems for the enhanced treatment of wastewater [101].

2.5 Electrochemical Advanced Oxidation Processes and its Mechanism

The detailed discussion on the mechanism and application of different electrochemical process like Direct oxidation, AEC, PC, and EF are included in this section.

2.5.1 Direct Oxidation

Direct oxidation or anodic oxidation in electrochemical wastewater treatment involves the direct transfer of electrons from the electrode (usually the anode) to the pollutant, while indirect oxidation relies on oxidizing species generated at the electrode, such as hydroxyl radicals or active chlorine, to degrade contaminants in solution [102,103].

Direct Electrochemical Oxidation

Direct electrochemical oxidation (also called anodic oxidation) operates by direct electron transfer between the pollutant adsorbed on the anode and the anode surface, without participation of additional chemical mediators [103]. The general steps involved are adsorption Eq. (2.6), electron transfer Eq. (2.7), and desorption Eq. (2.8) as shown below.

Adsorption of the pollutant (R) onto the anode surface (M) Eq. (2.6)

$$R + M \rightarrow M - R_{ads} \tag{2.6}$$

Electron transfer from the adsorbed pollutant to the anode Eq. (2.7)

$$M - R_{ads} + e^- \rightarrow M - R_{ads,ox} \quad (2.7)$$

Desorption of oxidized pollutant Eq. (2.8)

$$M - R_{ads,ox} \rightarrow R_{ox} \quad (2.8)$$

For many organic substances, the direct electron transfer results in partial or complete oxidation. The mechanism depends strongly on the nature of the anode material. Non-active anodes, such as Boron-Doped Diamond (BDD), tend to produce large amounts of surface-bound hydroxyl radicals, enabling complete mineralization of organic matter to CO_2. In contrast, active anodes, such as graphite, IrO_2, and Pt, favour selective (partial) oxidation due to their ability to stabilize certain intermediates. Accordingly, the production of surface hydroxyl radicals is shown in Eq. (2.9).

$$MO_x + H_2O \rightarrow MO_x(HO^\bullet) + H^+ + e^- \quad (2.9)$$

MO_x represents general oxide anode surface sites for the adsorption of $HO^\bullet$ Eq. (2.9). Depending on the electrode potential, these radicals may transfer oxygen to the lattice (chemisorbed active oxygen) as shown in Eq. (2.10).

$$MO_x(HO^\bullet) \rightarrow MO_{x+1} + H^+ + e^- \quad (2.10)$$

Subsequently, the substrate is adsorbed onto the surface of the active anode, where it undergoes oxidation, as represented in Eq. (2.11). In contrast, when the anode remains non-active, the organic substrate is directly mineralized through direct oxidation, as shown in Eq. (2.12).

$$MO_{x+1} + R \rightarrow MO_x + RO \quad (2.11)$$

$$MO_x(HO^\bullet) + R \rightarrow MO_x + CO_2 + H_2O + H^+ + e^- \quad (2.12)$$

Indirect Electrochemical Oxidation

In indirect oxidation, oxidizing agents or radicals are generated electrolytically at the electrode and subsequently react with contaminants in the bulk solution. The most important agents include hydroxyl radicals ($HO^\bullet$), active chlorine species (Cl_2, HOCl, OCl^-), and persulfate ($S_2O_8^{2-}$) [104]. The detailed discussion on the generation of hydroxyl radical with respect to the electrochemical process were discussed separately. The presence of excess concentration of chloride in the wastewater can electrochemically generate chlorine and subsequently to hypochlorous acid and hypochlorite ion as shown in Eqs. (2.13)-(2.15) [105,106]. The generated chlorine species especially reacts with the ammoniacal nitrogen present in the wastewater and oxidize the ammoniacal nitrogen to nitrogen as shown in Eq. (2.16) [107].

$$2Cl^- \rightarrow Cl_{2(g)} + 2e^- \quad (2.13)$$

$$Cl_2 + H_2O \rightarrow HOCl + H^+ + Cl^- \quad (2.14)$$

$$HOCl \rightleftharpoons OCl^- + H^+ \quad (2.15)$$

$$NH_4^+ + HOCl \rightarrow 0.5N_2 + 1.5\,H_2O + 2.5\,H^+ + 1.5\,Cl^- \quad (2.16)$$

The presence of sulphate ions in the wastewater can leads to the electrochemical generation of persulphate as shown in Eq. (2.17) [108]. Even though the persulphate generation in the electrochemical reactor was reported to be at a lower rate, the generated persulfate often activated by heat, ultraviolet light, or transition metals to produce sulfate radicals as shown in Eq. (2.18) [109].

$$2SO_4^{2-} \rightarrow S_2O_8^{2-} + 2e^- \quad (2.17)$$

$$S_2O_8^{2-} \rightarrow 2SO_4 \bullet^- \quad (2.18)$$

Sulfate radicals are strong oxidants which enable the breakdown of highly stable organic compounds, which oxidizes the recalcitrant organic matters.

2.5.2 Electro-Fenton, Peroxicoagulation, Electrochemical Peroxidation and Fered-Fenton Process

Electro-Fenton, Peroxicoagulation, Electrochemical Peroxidation (ECP), and Fered-Fenton (EF - Fere) are advanced electrochemical wastewater treatment methods based on Fenton chemistry, offering powerful pollutant degradation and broad industrial application [110].

Electro-Fenton

The EF process involves the in-situ generation of hydrogen peroxide (H_2O_2) at the cathode by reduction of dissolved oxygen. Fe^{2+} ions, added or generated via a sacrificial iron anode, catalyze the decomposition of H_2O_2 into highly reactive hydroxyl radicals ($HO^{\bullet}$), which attack and mineralize organic pollutants in acidic conditions [110].

At cathode, the cathodic reaction leads to the generation of hydrogen peroxide as shown in Eq. (2.19) [111]. And the hydrogen peroxide further gets reduced to hydroxyl radical by the Fe^{2+} oxidation to Fe^{3+} as shown in Eq. (2.1) [111]. The generation of hydrogen peroxide in the EF process depends on the pH of the wastewater and the acidic pH prefers the reaction.

$$O_2 + 2H^+ + 2e^- \rightarrow H_2O_2 \tag{2.19}$$

The anode may facilitate regeneration of Fe^{2+} or contribute by generating additional oxidants. This method is highly effective for degrading persistent organics, dyes, pharmaceuticals, and industrial pollutants, and it operates under mild conditions with high mineralization competence. EF achieved 84.3% COD removal in composite industrial wastewater, outperforming classical Fenton in terms of removal rate 78.43% and operating cost [112]. EF with carbon-felt cathode has demonstrated rapid and near-complete mineralization of pesticides, pharmaceuticals, and dyes in both lab and pilot trials [11].

Peroxicoagulation

Peroxicoagulation merges the principles of Fenton oxidation and electrocoagulation. Here, a sacrificial iron anode releases Fe^{2+} ions and simultaneously produces flocculants (mainly FeOOH, $Fe(OH)_3$), which aid in pollutant removal via adsorption and precipitation. Simultaneously, H_2O_2 supplied externally or generated in-situ reacts in Fenton fashion to yield hydroxyl radicals ($HO^{\bullet}$) [114,115].

Iron dissolution (anode) Eq. (2.20)

$$Fe \rightarrow Fe^{2+} + 2e^- \tag{2.20}$$

Fenton reaction (pollutant oxidation) Eq. (2.1)

Floc formation and pollutant entrapment occur as pH rises, linking coagulation and oxidation for enhanced mineralization and by-product removal [114,115]. PC has shown high mineralization rates of recalcitrant organic pollutants (pharmaceuticals, dyes) and significant improvement in wastewater biodegradability, with floc formation aiding in the removal of degradation by-products. The study explores an enhanced peroxicoagulation process for industrial wastewater treatment featuring iron anodes for coagulation and hydroxyl radical generation via peroxide production. Under optimized conditions of pH 3, the process achieved approximately 77.7% removal of COD and 97% removal of color [116]. The system effectively combined electrocoagulation and advanced oxidation to treat complex wastewater containing organic pollutants and nutrients. This synergistic coupling demonstrated the potential for improved contaminant removal efficiencies compared to individual treatments, highlighting peroxicoagulation's applicability for industrial wastewater reuse and discharge compliance [115].

Electrochemical Peroxidation

ECP is a variant of the classic Fenton treatment where Fe^{2+} ions are generated electrochemically from a sacrificial iron or steel anode and H_2O_2 is continuously supplied externally[3,110]. The process operates in undivided tank reactors, where the following reactions takes place on the anode and cathode surface.

Sacrificial iron / steel anode reaction Eq. (2.21),

$$Fe \rightarrow Fe^{2+} + 2e^- \tag{2.21}$$

Cathode reaction (hydrogen evolution) Eq. (2.22),

$$2H^+ + 2e^- \rightarrow H_2 \uparrow \tag{2.22}$$

In the bulk solution the Fenton reaction or the main pollutant attack reactions takes place as given in Eq. (2.1).

$Fe(OH)_3$ floc may also assist via coagulation in the same reactor if the pH condition of the favours and the process enables excellent removal of Volatile Organic Compounds (VOCs), Polychlorinated Biphenyl (PCBs), heavy metals, pesticides, dyes, and real complex wastewaters [117]. A study on electrochemical peroxidation for landfill leachate treatment showed 91% COD removal and 95% color reduction under optimized conditions of pH 3, H_2O_2 dosage of 3 mL/L, Fe^{2+} of 1.5 g/L, 6 V applied voltage, and Fe-Al electrode pair over 120 minutes of [118]. The study highlighted not only high pollutant removal but also a marked improvement in effluent biodegradability and energy efficiency with energy consumption as low as 3.6 kWh/m^3 of wastewater treatment.

These results demonstrate the effectiveness of ECP for mineralizing organic and inorganic contaminants, transforming landfill leachate into a more biodegradable effluent suitable for further treatment or discharge [118].

Fered-Fenton Process

The Fered-Fenton process employs an undivided electrolytic cell where both Fe^{2+} or Fe^{3+} and H_2O_2 are externally added, and the cathode reduces Fe^{3+} back to Fe^{2+} [110,119].The Fe^{3+} reduction reaction on the cathode surface is as given in Eq. (2.23).

$$Fe^{2+} + e^{-} \rightarrow Fe^{3+} \tag{2.23}$$

Fenton reaction (main pollutant attack) Eq. (2.1) and cathode reaction (hydrogen evolution) Eq. (2.22).

The design of the electrochemical reactor ensures continuous Fe^{2+} regeneration and minimizes $Fe(OH)_3$ precipitation at optimized low pH values. The process is distinguished by higher iron concentrations, serving as a reactant, not a catalyst. In a benchmark study treating benzene sulfonic acid at pH 2, Fered-Fenton achieved a COD removal of 87% after 90 minutes of electrolysis, while chemical Fenton reached only 73.5% COD removal in the same period [119]. This improved performance stems from the process design; Fe^{3+} in solution is reduced back to Fe^{2+} at the cathode, maintaining optimum catalyst content for efficient hydroxyl radical ($HO^{\bullet}$) generation via the Fenton reaction. The enhanced mineralization effect is attributed to this uninterrupted Fe^{2+} supply, which drives greater·$HO^{\bullet}$ production and sustained oxidation power even as the reaction progresses [119].

2.5.3 Key Parameters Influencing Electrochemical Advanced Oxidation Process Performance

Several operational and water quality parameters critically impact the efficacy and efficiency of electrochemical advanced oxidation process. The electrode materials, current density and applied voltage, pH, supporting electrolyte and conductivity, pollutant type and concentration, electrolysis time, electrode spacing electrode area, depth of exposure of the electrode and reactor design, etc., were the major factors which influences the electrochemical process efficiency. The detailed impact of each and every factors or parameter which influences on the different electrochemical process are discussed below.

Electrode Material

The electrode material is a fundamental parameter affecting the performance of EAOPs. The selection of electrode materials determines on the type and amount of reactive species to be generated, directly impacting the mineralization efficiency of complex pollutants.

Sacrificial anodes, commonly iron [120] dissolve during the treatment process, releasing metal ions which promote coagulation and facilitate Fenton-like reactions. This dual mechanism of pollutant removal oxidation by hydroxyl radicals and coagulation via metal hydroxides enhances overall treatment efficacy. Nayir et al. [121] reported a case study on laundry wastewater where an iron anode combined with a carbon cathode achieved 75% COD removal under optimal current density 45 mA/cm^2 and pH 7 conditions. The study highlighted the practical effectiveness of such sacrificial electrodes in real wastewater matrices while maintaining residual iron concentrations low enough (0.9 mg/L) to adhere to environmental discharge limits. The approach illustrates the synergy between electrochemical oxidation and coagulation that sacrificial anodes offer in peroxicoagulation systems, making them suitable for treating wastewaters rich in suspended solids and organic pollutants.

Non-active anodes such as BDD, lead dioxide (PbO_2), tin dioxide (SnO_2), and Magnéli phase titanium suboxides (Ti_4O_7) are characterized by their high oxygen evolution potentials. Unlike active anodes, they facilitate the generation of physisorbed hydroxyl radicals that promote near-complete mineralization of organic contaminants with minimal formation of intermediate products. El Kateb et al. [122] demonstrated that treatment of landfill leachate using Ti_4O_7 anodes combined with Fe(II/III) layered double hydroxide-modified carbon felt cathodes achieved up to 96% removal of dissolved organic carbon (DOC). This study confirms the exceptional oxidation capability of Ti_4O_7. Panizza and Martinez-Huitle [123] carried out a study for the treatment of landfill leachate using Ti–Ru–Sn ternary oxide ($TiRuSnO_2$), PbO_2 and BDD anodes and stainless steel as cathode. The experimental results demonstrated that following 8 hours of electrolysis, the $TiRuSnO_2$ anode achieved only 35% COD removal, 52% color removal, and 65% ammonium removal. The application of PbO_2 resulted in the complete removal of color and ammonium; however, a residual COD remained. In contrast, BDD facilitates comprehensive COD, color, and ammonium elimination.

Active anodes such as graphite, platinum, and mixed metal oxides like IrO_2, RuO_2, etc., which were less oxidative than BDD, provide partial oxidation mechanisms useful for pre-treatment or less refractory wastewaters. These materials facilitate catalytic oxygen group formations that oxidize pollutants, balancing efficiency and electrode durability. These active anodes for recalcitrant industrial wastewaters, benefiting from low electrode deterioration and high current efficiency. Periyasamy & Muthuchamy [124] conducted a comprehensive study using a graphite anode for the electrochemical treatment of paracetamol in aqueous solution, focusing on removal efficiencies of paracetamol, COD, and TOC, along with the effect of key operational parameters. Under optimized conditions pH 7, sodium chloride as supporting electrolyte at a concentration of 0.2 M, and a current density of 40 mA/cm^2, the authors achieved a paracetamol removal efficiency of 98%, COD removal of 82%, and TOC reduction of 63% after 120 minutes of electrolysis.

Current Density and Applied Voltage

Current density and applied voltage are pivotal parameters in electrochemical advanced oxidation processes, as they govern the generation rate of reactive oxidant species and influence process energy consumption. Current density directly correlates with the production of hydroxyl radicals and other oxidative species on the anode surface [125, 126]. Increasing current density generally accelerates the mineralization of organic pollutants by producing greater hydroxyl radical flux, enhancing degradation rates. However, beyond an optimal threshold, further increases lead to parasitic reactions such as oxygen evolution, reducing current efficiency and wasting energy [127]. Najafinejad et al. [103] demonstrated in pharmaceutical wastewater treatment that while elevating current density improved pollutant removal, energy consumption became disproportionately high after exceeding the optimal value, of 15 mA/cm^2 adversely impacting process sustainability. Applied voltage influences the anode potential, controlling the oxidation power. Controlling voltage is crucial because an excessively high voltage can cause rapid electrode degradation, increased side reactions, and energy losses [127]. Most EAOP systems achieve an operational balance within a voltage range of approximately 5 to 10 V for the lab studies, optimizing treatment efficiency while minimizing costs [125-127]. This balance ensures adequate radical generation for effective contaminant mineralization without excessive energy expenditures or damage to electrode materials.

pH

The efficiency of the electro-Fenton and Peroxicoagulation process critically dependent on maintaining the solution pH in the acidic range, typically between 2.5 and 3.5, as this provides the optimal environment for the catalytic activity of Fe^{2+} ions and the generation of hydroxyl radicals ($HO^{\bullet}$). At this pH, Fe^{2+} remains soluble and highly reactive with hydrogen peroxide, ensuring sustained radical production through the Fenton reaction[110,128]. When the pH rises above 3.5–4.0, ferric iron (Fe^{3+}) starts to hydrolyze and precipitate as $Fe(OH)_3$, reducing the availability of dissolved Fe^{2+}/ Fe^{3+} for the catalytic cycle and causing a major drop in radical yield and process efficiency. At the same time, very low pH (< 2.0) reduces process efficiency by stabilizing less-reactive iron or peroxide complexes and enhancing side reactions that consume H_2O_2 without producing radicals [129].

Jegadeesan et al. [118] treated landfill leachate using the electro-Fenton process and achieved over 90% COD removal when the pH was maintained at 3.0. Deviating from this optimum, especially toward higher pH, led to iron precipitation and significant reductions in pollutant removal due to diminished hydroxyl radical production. This finding is corroborated across multiple electro-Fenton applications, including synthetic and real wastewaters, where a pH around 3 always ensures maximum process efficiency and stability of iron catalysts. Venu et al. [130] treated stabilized landfill leachate using the peroxicoagulation process and achieved over 93% COD removal when the initial pH was maintained at 3.0, voltage of 10 V and reaction time of 90 min.

Supporting Electrolyte and Conductivity

Supporting electrolytes such as sodium sulfate (Na_2SO_4), sodium chloride (NaCl), and sodium nitrate ($NaNO_3$) are frequently employed in EAOP to enhance solution conductivity, thus reducing cell resistance and energy consumption. Sodium sulfate is the most commonly used electrolyte because it is inert and does not generate potentially harmful byproducts. Studies have shown that Na_2SO_4 enhances the production of hydroxyl radicals and promotes better degradation rates of contaminants [131,132].

Wang et al. [133] demonstrated improved degradation of perfluorooctane sulfonate (PFOS) using Ti_4O_7 electrodes in the presence of Na_2SO_4, achieving up to 98.6% removal. In contrast, chloride-containing electrolytes (NaCl) can lead to the generation of secondary oxidants such as chlorine species (Cl_2, HOCl), which can enhance pollutant degradation but also bring about the formation of disinfection byproducts like chlorinated organics, posing environmental risks. If the electrolyte is carbonate medium the carbonate species will react with $HO^{\bullet}$ resulted in the scavenging reactions and the subsequent formation of weak carbonate radical species. Compare to other AOPs, EAOPs have extra sources of carbonate radical species during the electrochemical oxidation of solution containing carbonate species [134].

Pollutant Type and Concentration:

The pollutant type and concentration play crucial roles in the electrochemical advanced oxidation process performance by influencing degradation kinetics and mechanisms [135]. In general, more reactive organic compounds such as phenols and antibiotics degrade faster [136,137], whereas halogenated aromatic compounds exhibit slower degradation due to their chemical stability and resistance to oxidative attack [138]. Studies have shown that as pollutant concentration increases, oxidation rates generally decrease because the available reactive species, such as hydroxyl radicals, must be shared among more pollutant molecules, leading to reduced degradation rate constants and efficiencies[139].

According to Khan et al. [140] an increase in initial pollutant concentration generally leads to a decrease in oxidation rate constants and overall removal efficiency due to the limited availability of reactive oxidizing species such as hydroxyl radicals. This phenomenon occurs because higher contaminant loads impose increased competition for the radicals and active sites on the electrode surface, promoting scavenging interactions that reduce the effective radical concentration available for target molecule destruction. In another study, Panizza & Cerisola [125] highlighted that the electrochemical treatment of complex pharmaceutical wastewater with mixed pollutant loads requires longer treatment times and optimal operating conditions to achieve effective mineralization.

Electrolysis Time

Longer electrolysis times generally facilitate higher pollutant removal efficiencies as they provide extended opportunities for the generation of hydroxyl radicals ($HO^{\bullet}$) and other reactive oxygen species to degrade target contaminants. However, treatment duration must be optimized to balance energy consumption, process throughput, and operational costs [110]. Excessive electrolysis times can lead to diminishing returns where removal efficiencies plateau, while also increasing energy demand and accelerating electrode wear, reducing system longevity [141].

Wang et al. [3] demonstrated that optimal electrolysis time of 240 min in the sequential ECP-EF process enabled effective organic pollutant removal with minimized acute cytotoxicity and energy expenditure. The study highlights that electrolysis times beyond the optimum range do not provide significant additional removal but increase costs, thus suggesting economic and environmental benefits of treatment time optimization.

Electrode Spacing and Reactor Design

Electrode spacing and reactor design are critical factors in optimizing electrochemical advanced oxidation processes, impacting energy efficiency and treatment performance.

Reduced electrode spacing decreases ohmic resistance within the reactor, leading to lower energy consumption and improved current efficiency [142]. Normally the electrode spacing ranges between 0.5 cm to 4 cm [143-146]. El Kateb et al. [122] exemplified this principle by designing a Ti_4O_7 based electrochemical reactor with closely spaced electrodes for landfill leachate treatment. The compact electrode arrangement enhanced mass transport and minimized voltage losses, contributing to a high DOC removal efficiency of 96% with favorable energy consumption metrics. Senguttuvan et al. [147] varied electrode spacing between 2 and 6 cm and observed that COD removal efficiency (25.9%) was decreased, when electrode spacing was increased from 4 to 6 cm, where COD removal efficiency (64.3%) was increased when electrode spacing was decreased between electrodes (from 4 cm to 2.5 cm). Thus, about 2.5 cm of distance between the electrodes was the optimal electrode spacing for accelerated pollutant degradation in distillery wastewater by the EF process.

2.5.4 Electrochemical Advanced Oxidation Process Applications in Real Wastewater Treatment

Electrochemical advanced oxidation processes have found widespread application in the treatment of real wastewaters from diverse industrial and municipal sources, owing to their ability to degrade recalcitrant contaminants that persist after conventional biological and physicochemical steps.

EAOPs such as anodic oxidation, electro-Fenton, heterogeneous EF, Peroxicoagulation etc., have been implemented to address the challenge of treating landfill leachate, industrial, municipal, and pharmaceutical wastewaters containing non-biodegradable pollutants [103,133,139]. The use of heterogeneous electro-Fenton with Fe(II/III)-Layered Double Hydroxide (LDH)-modified carbon felt and Ti_4O_7 anodes for nanofiltration concentrate (NFC) from landfill leachate pretreated by membrane bioreactor. Here, 96% dissolved organic carbon removal was achieved after 8 hours of electrolysis at circumneutral pH and 8.3 mA/cm^2, highlighting the potential of such EAOPs to mineralize highly refractory organic matter in actual complex effluents. When combined with Ti_4O_7 anodic oxidation at lower current densities, the process achieved substantial removal (77% DOC) at lower energy cost (0.13–0.11 kWh/g DOC removed), making it attractive for industrial adoption. Biodegradability indices and acute toxicity assays further confirmed a reduction in toxicity and an increase in the biological amenability of the treated effluent [122].

EAOPs coupled with biological treatments have achieved marked removal of color, COD, and acute toxicity in textile and dye-industry wastewaters that typically resist aerobic degradation. For example, electrochemical oxidation with boron-doped diamond and SnO_2 anodes has shown near-complete color removal. Bravo Yumi et al. [148] used Ti/IrO_2-SnO_2-Sb_2O_2 anode to remove textile azo dye and achieved a color and COD removal of 100% in 20 min and 90% in 60 min at a current density of 50 mA/cm^2. The electrochemical oxidation of textile effluents over a boron-doped diamond anode achieved a complete decolorization accompanied by significant mineralization up to 85%

with an optimum condition of current density of 8 mA/cm^2, pH 1 and 180 min of reaction time [149]. While integrated photo electro-Fenton processes are commonly used for high-strength dye mixtures and pharmaceutical contaminants. Integrated photo electron-Fenton (BDD/CF, UVA/Solar) attains 100% COD removal on high strength dye mixture at an operating condition of 20 mA/cm^2, [Fe^{2+}] of 0.1 mM, pH 3 and reaction time of 120 min [150]. Integrated photo electron-Fenton (RuO_2/ACF, UVA light) attains 100% Cefalexin removal at an operating condition of 360 mA, pH 3 and reaction time of 360 min [151].

A study involving peroxicoagulation for real wastewater treatment in the removal of pollutants from laundry wastewater, the process employed iron anodes and carbon-polytetrafluoroethylene cathodes and reached optimal treatment conditions at pH 7, 45 mA/cm^2 current density, and 25 °C. Under these conditions, the peroxicoagulation process achieved 75% COD removal, 67% methylene blue active substances (MBAS, a proxy for surfactants) removal, and nearly 100% total phosphorus removal. Additionally, the process minimized residual iron to just 0.9 mg/L, making it viable for reuse scenarios. The mechanism combined in-situ generation of H_2O_2 (by the cathode) and persulfate, which together with iron precipitation and radical oxidation enabled robust removal of organic and nutrient contaminants [121].

Studies on petrochemical wastewater found that direct anodic oxidation using mixed metal oxide and parallel plate reactors efficiently reduced COD levels to 20,450 mg/L from 52,300 mg/L with significant reductions of COD about 79% [152]. For pharmaceutical and hospital effluents, anodic oxidation and EF have shown strong removal (> 90%) of antibiotics, hormones, and priority micro-pollutants, while also addressing transformation product concerns [103].

2.6 Aerated Electrocoagulation and Peroxicoagulation

The detailed discussion on the mechanism of electrocoagulation, aerated electrocoagulation, peroxicoagulation and sequential electrochemical advanced oxidation process combinations are included in this section.

2.6.1 Mechanism of Electrocoagulation: iron/aluminum dissolution, floc formation

Electrocoagulation is an electrochemical wastewater treatment technique where sacrificial metal electrodes typically iron or aluminum are dissolved electrochemically to generate coagulant species in situ as shown in Fig 2.1. At the anode, iron or aluminum dissolves, releasing Fe^{2+} or Al^{3+} ions Eq. (2.21) [153]. These ions rapidly hydrolyze to form metal hydroxides ($Fe(OH)_2$, $Fe(OH)_3$, $Al(OH)_3$), which act as destabilizing and adsorbing agents for colloidal particles and dissolved contaminants Eqs. (2.24-2.27) [154]. The positive metal hydroxide flocs neutralize the negative charges of suspended particles. The formed amorphous metallic hydroxides have large surface areas with very high adsorption capacity, make bond with the pollutants and forms flocs. The metallic flocs enlarge the size from micro flocs to macro flocs attaining the self-settling size Eq. (2.33) [155]. This eliminates the need for dosing with external chemical coagulants, reduces sludge production, and is efficient for removing colloids, heavy metals, oil, organic pollutants, and dyes [156–159].

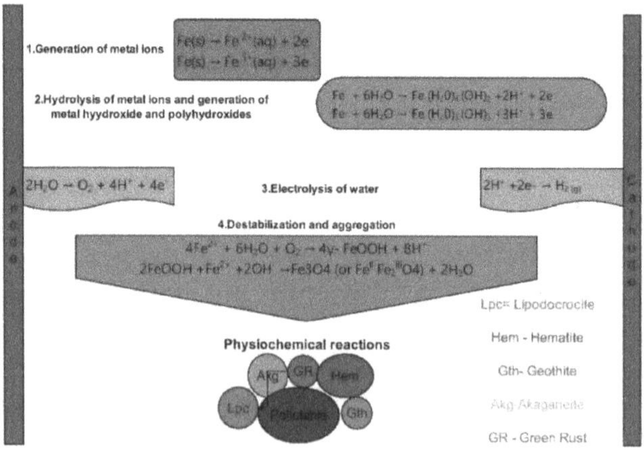

Fig. 2.1 Diagrammatic representation of floc generation and pollutant removal mechanism during iron EC process

$$Fe^{2+} + 2OH^- \rightarrow Fe(OH)_2 \quad (2.24)$$

$$Fe^{3+} + 3OH^- \rightarrow Fe(OH)_3 \quad (2.25)$$

$$4Fe(OH)_2 + 2H_2O + O_2 \rightarrow 4Fe(OH)_3 \quad (2.26)$$

$$\text{Pollutant} + Fe(OH)_3 \rightarrow \text{Precipitate} \quad (2.27)$$

In long-term field operation, EC electrodes often develop macroscopic passivation layers dominated by Fe_3O_4 (magnetite). These layers trap freshly generated Fe(II) near the electrode, hindering its transport into the bulk where pollutant removal occurs; the result is lower coulombic performance [160]. The growth of thick magnetite layers on EC electrodes has been specifically documented as a treatment challenge, with mineralogical evidence showing millimeter-scale deposits that accumulate under typical on / off polarization and wet / dry cycles [29,160]. EC mineral phases formed under different redox regimes (e.g., magnetite and green rust in low DO conditions) can also evolve over time, complicating process control and separation of precipitates [27].

2.6.2 Aerated Electrocoagulation

In aerated electrocoagulation, the iron anode still supplies Fe(II) / Fe(III) that hydrolyze to form voluminous $Fe(OH)_x$ "sweep-flocs," so separation by adsorption, charge neutralization and enmeshment remains the dominant removal pathway, just as in conventional EC. The difference is that continuous aeration rapidly converts dissolved Fe(II) to Fe(III) in the bulk [29] as shown in Eq. (2.28)

$$2Fe^{2+} + \frac{1}{2}O_2 + 2H^+ \rightarrow 2Fe^{3+} + H_2O \quad (2.28)$$

sustaining a steady supply of ferric hydroxide flocs the principal coagulating species in iron EC until dissolved oxygen is depleted [27]. Aeration also improves the coagulation environment bubble-induced turbulence and extra gas liquid interfacial area increase mass transfer of freshly generated coagulant, lengthen and intensify coagulant–colloid contact, and can raise particle collection efficiency [161].

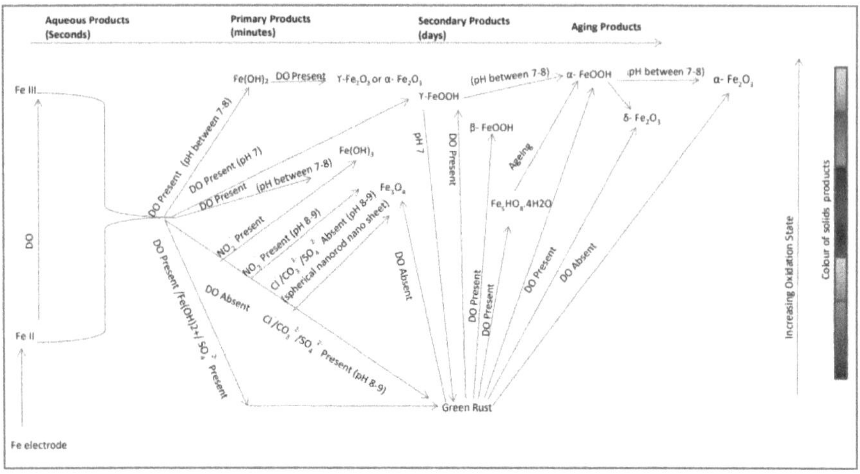

Fig. 2.2: Development and production of flocs from Fe(0) EC process by the effect of solution chemistry

Aeration also helps manage anode passivation, a key durability limitation in field EC. when dissolved O_2 is present and mixed into the electrolyte as in AEC mixed-valent iron phases (magnetite / green rust) that cement onto the electrode form only when DO is absent or very low (Eq. (2.29) and (2.30)) [27]; under oxygenated conditions, ferric oxyhydroxides dominate instead of other species as shown in Eq. (2.31). This shift matters for passivation: fast, bulk-phase oxidation of Fe(II) pulls iron away from the electrode surface (shortening its residence time in the stagnant boundary layer where magnetite films grow) and promotes precipitation in suspension as removable Fe(III) floc rather than as adherent films [27]. Fig. 2.2 summarizes the various circumstances and pathways for mixed valent Fe flocs production with and without oxygen.

$$2FeOOH + Fe^{2+} + 2OH^- \rightarrow Fe_3O_4\,(\text{or } Fe^{II}Fe_2^{III}O_4) + 2H_2O \quad (2.29)$$

$$4Fe(OH)_2 + 2\gamma - FeOOH + SO_4^{2+} 6OH^- + 2H_2O \rightarrow Fe_3O_4\,(\text{or})$$

$$Fe_4^{II}Fe_2^{III}(OH)_{12}SO_4 \quad (2.30)$$

$$4Fe^{2+} + 6H_2O + O_2 \rightarrow 4\gamma - FeOOH + 8H^+ \quad (2.31)$$

2.6.3 Peroxicoagulation

Peroxicoagulation enhances EC process by additionally generating hydrogen peroxide (H_2O_2) in situ. At the cathode, dissolved oxygen is electrochemically reduced to H_2O_2 Eq. (2.19) [116].

Iron (II) generated at the anode reacts with H_2O_2 (a Fenton-like process), producing reactive hydroxyl radicals ($HO^{\bullet}$) with strong oxidizing capacity Eq. (2.1) at the acidic pH [116].

Thus, PC combines the physical removal of pollutants by coagulation / flocculation with oxidative degradation via·$HO^{\bullet}$ radicals, broadening the spectrum of treatable contaminants and improving removal rates of refractory organics [142]. The study by Venu et al. [130] investigated the treatment of stabilized landfill leachate using the peroxicoagulation process in a batch electrolytic reactor equipped with stainless steel anodes and graphite cathodes. The research explored the effects of various operating parameters including pH, applied voltage, electrode area, inter-electrode distance, and hydrogen peroxide addition on the process efficiency. The authors found that controlling the pH at an optimal level of 3 significantly enhanced the production of hydroxyl radicals and improved the solubility of iron species, which are essential for effective oxidation and coagulation mechanisms in PC process. Under optimized conditions, such as a voltage of 10 V, electrode spacing of 4 cm, electrode area of 25 cm^2, and a solution depth of 6.8 cm the process achieved a high COD removal efficiency of approximately 91% within 120 minutes. The study also highlighted that stepwise addition of hydrogen peroxide was more effective than a single batch addition by minimizing the scavenging of hydroxyl radicals. Furthermore, iron anodes outperformed stainless steel in accelerating COD degradation, albeit with greater sludge production. Increase in electrode surface area and depth of immersion of cathode enhanced the generation of reactive species, further improving organic pollutant mineralization. This case study demonstrates that the peroxicoagulation process can be effectively applied to landfill leachate treatment, offering a controllable, environmentally friendly approach with high removal efficiencies under proper operational conditions [130].

2.6.4 Sequential Electrochemical Advanced Oxidation Process Combinations

Recent advances in leachate and complex wastewater treatment focus on integrating multiple electrochemical processes in sequence to exploit cumulative and synergistic effects. New research demonstrates much higher efficiencies by employing hybrid, staged approaches each tailored to tackle specific pollutant classes and wastewater composition challenges [3].

A two-step electrochemical method ECP followed by electro-Fenton was evaluated on real old landfill leachate. At current density of 30 mA/cm^2, ECP, pH 9.0 and 30 mA/cm^2, EF, pH 3.5, the process removed 95.9 ± 0.4% COD, 85.1 ± 0.8% TOC, 99.4 ± 0.1% NH_4–N, and 99.2 ± 0.1% TN. Beyond high removals, spectroscopic tracking showed a marked loss of aromatic humic character and a decline in acute cytotoxicity, consistent with reduced molecular complexity after ECP + EF. For the economical configuration with a detention time of 240 min, the treatment cost was $0.0134 per liter of wastewater, which was 32% lower than a previously reported electrocoagulation + electro-Fenton sequence, while achieving 36.9% higher COD removal and 34.5% higher NH_4^+–N removal. This synergy, together with the demonstrated toxicity reduction, makes the ECP + EF method a strong candidate as an advanced pretreatment ahead of biological polishing treatment [3].

A three-stage treatment study by electrocoagulation with aeration (Al electrodes), followed by persulfate-based oxidation (PMS / ZVI / UV), and then electro-Fenton (Carbon Felt and Pt electrodes) was applied to real, mature landfill leachate and delivered stepwise treatment with clear synergy. Starting from a raw COD of 4,040 mg/L, the AEC pretreatment removed 61% COD, after which PMS / ZVI / UV pushed COD down to 300 mg/L, and the final EF stage achieved 80 mg/L, in parallel, NH_4–N fell from 367 to 26 mg/L (overall 93% removal). Across the sequence the biodegradability index increased markedly (BOD_5/COD from 0.18 to 0.56), and phytotoxicity abated to non-toxic levels (germination index > 90% for tomato and radish; garden cress 78.8%), consistent with the concurrent collapse of aromatic character ($SUVA_{254}$ from 0.45 in the raw leachate to 0.13 after EF). Mechanistically, the authors show that 5.2 mM of PMS carried over from the PMS / ZVI / UV step is further activated cathodically during EF (to HO• and $SO_4^{\bullet -}$), reinforcing oxidation in the third stage, EF current optimization also mattered, with COD removal rising to 83% at 110 mA and declining at higher current due to side reactions underscoring why the EF block was positioned last and run at its own optimum [162].

Sequential electrochemical advanced oxidation systems applied to complex industrial wastewaters showed that the coupled process not only improved pollutant removal but also reduced treatment time and energy requirements compared to single-stage operation.

2.7 Electrode Materials for Electrochemical Advanced Oxidation Process

In Electrochemical Advanced Oxidation Processes, electrode materials govern the efficiency and sustainability of contaminant degradation pathways. Both anode and cathode play critical roles in generating oxidative species like hydroxyl radicals and hydrogen peroxide (H_2O_2) that drive mineralization of persistent compounds in industrial wastewater and stabilized landfill leachate [34]. The electrochemical activity, chemical durability, conductivity and mechanical strength of electrode materials ultimately determine treatment performance, cost-effectiveness, and scalability [34]. The $2e^-$ oxygen reduction process (ORR) is a key factor in the EAOP's efficiency, which influences the cathode H_2O_2 production. The cathode material and the efficiency of its H_2O_2 electro generation greatly influence the cost and efficacy of the process. The challenges with the use of different anodes and cathodes in the EAOP's are given in Table 2.2.

2.7.1 Cathode Materials: Conventional and Emerging Options

Carbonaceous materials, including activated carbon fibre, graphite, graphite felt etc have been identified as promising cathode materials for the electrolytic generation of H_2O_2. This is due to their presence in abundance, its high electrical conductivity, the chemical stability, mechanical robustness and low cost. However, due to the complications involved in fabrication and the need for enhanced H_2O_2 generation, researchers have explored other effective cathode materials, which are briefly discussed below.

Conventional Carbon-Based Cathodes

Cathodes made from graphite, carbon felt, or carbon cloth are extensively used for H_2O_2 generation through the two-electron ORR in acidic environments [163]. Their combination of high surface area, electrical conductivity, and cost-efficiency makes them suitable for electro-Fenton and peroxicoagulation applications [164]. Even though carbon materials like graphene, carbon nanotubes, carbon nanofibers etc shows desirable mechanical and electrical properties, the fabrication process is still complex, and the initial raw materials typically consist of fossil resources that are non-renewable [22].

Over time, though, fouling, passivation, and loss of performance pose major limitations. This results in higher maintenance demand and inconsistent H_2O_2 generation, particularly under continuous or real-wastewater operation.

Biomass-Derived Carbon Cathodes

Recent studies have demonstrated that porous carbon derived from biomass, including agricultural residues and aquatic biomass such as water hyacinth, are promising renewable cathode materials for electrochemical advanced oxidation processes, particularly in electro-Fenton and ORR applications [165,166]. These biomass-derived carbons exhibit high porosity, abundant surface defects, and, when doped with heteroatoms such as nitrogen (N) and sulfur (S), significantly enhance ORR kinetics and the in-situ generation of hydrogen peroxide compared to traditional carbon felt electrodes [167,168].

For instance, nitrogen-doped biochar derived from water hyacinth showed a high specific surface area of 829 m^2/g and achieved approximately 1.7 mM/L H_2O_2 yield with 81% current efficiency during electro-Fenton processes. This translated to an 86.8% TOC removal in 30 minutes, substantially outperforming conventional graphite cathodes [165]. Similarly, biomass-derived carbon materials doped with sulfur and nitrogen from waste leaf biomass demonstrated efficient catalytic generation of reactive oxygen species (ROS), including hydroxyl radicals, singlet oxygen ($^1O_2$), and H_2O_2 across a wide pH range (3–11), maintaining excellent stability over multiple cycles [166]. The enhanced catalytic performance of these materials is largely attributed to the increased density of electroactive sites, notably pyridinic-N and graphitic-N, which improve electron transfer pathways due to optimized pore structures and tailored chemical composition [166,167].

Table 2.2 Comparative Analysis and Key Challenges of Different Types of Anodes and Cathode

Material	Role	Merits	Challenges	References
BDD	Anode	High $HO^{\bullet}$ yield, exceptional stability	High cost, energy-intensive production	[169–171]
Ti/MMO	Anode	Robust, scalable, moderate cost	Noble metal reliance, coating delamination risk	[172-175]
Graphite / Carbon Felt	Both	Low cost, accessible	Corrosion, fouling, reduced long-term performance	[176]
Biomass Carbon	Cathode	High H_2O_2 yield, sustainable, fouling-resistant	Requires standardization and long-term field validation	[177]
Stainless Steel	Cathode	Cheap, robust	Low catalytic activity, corrosion-prone in acidic media	[178].

From a sustainability viewpoint, biomass-derived cathodes possess a significantly lower environmental footprint than conventional cathode materials. These electrodes are manufactured from inexpensive, abundant, and renewable biomass sources, thus aligning with circular economy principles [168]. Furthermore, being metal-free, biodegradable, or recyclable, they substantially mitigate the environmental impacts traditionally associated with carbon felt and metal-based cathodes [165,167]. The combination of superior catalytic efficiency and environmental sustainability positions biomass-derived heteroatom-doped carbons as ideal candidates for scalable, environmentally friendly treatment of recalcitrant wastewater [166,168].

Electrode materials, particularly their electrochemical, economic, and environmental characteristics are central to the performance and feasibility of EAOPs. While BDD and Ti/MMO remain the most effective anodes, their economic and sustainability shortcomings compel the exploration of improved alternatives. Simultaneously, biomass-derived carbon cathodes illustrate great promise for sustainable, high-performance electrochemical treatment.

Table 2.2, shows the comparative analysis and key challenges of different types of anodes and cathodes. These, innovations offer impactful avenues for future landfill leachate and wastewater treatment system designs that are both cost-effective and environmentally responsible.

2.8 Biomass-Derived Carbon Electrodes

Recent work has explored a variety of biomass wastes as carbon precursors for EAOP cathodes. Aquatic weeds like water hyacinth yield N-rich biochar via molten-salt carbonization, producing hierarchically porous carbons without external dopants. bamboo stalks, pyrolyzed at 1000 °C and acid-washed, form graphitic biochar with favorable conductivity and surface chemistry for dual H_2O_2 generation or activation [179]. Algal biomass (blue-green algae) can be KOH-activated at 600 °C to self-dope nitrogen species (pyridinic, pyrrolic, graphitic N), enhancing ORR sites for H_2O_2 electrogeneration [180]. Leguminous wastes like long bean stems deliver carbons with superhigh surface areas up to 2270 m^2/g and abundant meso and macro-porosity [181]. Lastly, floral residues (magnolia flowers) produce mesoporous carbons (1226 m^2/g) upon direct pyrolysis at 800 °C [182].

2.8.1 Electrode Fabrication from Biomass

Water Hyacinth: $ZnCl_2$ molten-salt pyrolysis at 700 °C simultaneously carbonizes and self-dopes water hyacinth, yielding biochar powder with 829 m^2/g surface area and 43.2% pyridinic-N mixed with 10 wt% acetylene and 10 wt% PTFE. This mixture slurry was mixed with ethyl alcohol and finally suppressed evenly onto titanium mesh [165].

Bamboo: Raw bamboo is pyrolyzed at 1000 °C under N_2, cooled to 300 °C, HCl-washed, then assembled into graphitic bamboo biochar and a stainless steel mesh (GBSS) composites by wrapping graphitic biochar granules within stainless-steel mesh, binder-free and scalable [179].

Algal Biochar: Pre-carbonization at 400 °C followed by KOH activation and secondary pyrolysis at 600 °C, 700 °C and 800 °C produces activated porous biochar. This biochar sample was dispersed in an ultrasonic bath in polyvinylidene fluoride (PVDF) and then coated onto SS mesh to fabricate cathodes [180].

Long Bean: Long bean was washed and mixed with $KHCO_3$ and then pyrolysis at 800 °C for 2 h. The obtained material was washed with HCl and dried in oven at 105 °C then used directly as cathode catalysts in Triboelectric nanogenerator (TENG) driven EF systems [181].

Magnolia Flowers: Magnolia petals were pyrolyzed at 800 °C in N_2 to yield biochar material, then acid-washed and optionally $ZnCl_2$-activated (mass ratios 1:3–1:5) to tailor porosity and N content and dried in oven at 105 °C before TENG assembly [182].

Electrochemical Properties and Performance

Water hyacinth biochar sample delivers 1.7 mM/L H_2O_2 at 0.7 V with 81.2 ± 2.5% current efficiency, rivalling or exceeding the capacity of earlier reported biomass carbons [165]. Bamboo GBSS electrodes generate 11.3 mg /L H_2O_2 at 50 mA (neutral pH), boosted to 18.3 mg/L after O-doping, and activate H_2O_2 to $HO^•$ in Fe-free EF [179]. N,S-codoped wood biochar co-activates H_2O_2 and Fe^{2+} to produce $HO^•$, $^1O_2$, and $SO_4^{\cdot -}$, outperforming N-doped analogues with k_{obs} × 2.25 higher for sulfonamide degradation at mild pH [167]. Algal-derived biochar features enriched micropores conducive to O_2 access and H_2O_2 release, while activated carbons show BET areas around 800 – 1200 m^2/g depending on pyrolysis temperature [180]. Long bean carbons achieve up to 2270 m^2/g surface area with I_D/I_G as 0.88, correlating with superior ORR kinetics [181]. Magnolia-flower carbons exhibit 1226 m^2/g and I_D/I_G as 0.90, supporting high EF degradation rates under self-powered TENG conditions [182]. Table 2.3, shows the previous studies on the different biomass derived cathodes used as the electrode in the EF and PC process.

Table 2.3 Studies Using Biomass-Derived Cathodes in EF or PC Process

Biomass source	**Pollutant**	**Removal**	**Time**	**Reference**
Water hyacinth	Dimethyl phthalate	86.8% TOC removal	30 min	[165]
Algal biochar	Sulfadiazine	>96% degradation	4 h	[180]
Long bean	4-Dimethylaminoazo benzene	100% TOC removal	70 min	[181]
Magnolia flowers	Basic orange 2	98% degradation	70 min	[182]

2.9 . Research Gaps

- Limited studies used real landfill leachate: Most EAOP reports test synthetic model pollutants (e.g., sulfathiazole, sulfamerazine, basic dyes), not real leachate, so results omit the effects of refractory organics, NH_4–N, metals, and high salinity.
- Underexplored sequential EC-PC combinations for leachate: While EC + EAOP shows synergy in dye / textile effluents (e.g., 93.5% COD removal, lower sludge) analogous trials on leachate are scarce.
- Lack of performance validation for biochar cathodes: Despite the promising H_2O_2 yields, the studies have not focused on converting solid biomass into cathode electrodes through a single-step pyrolysis process, without any additional steps or chemical additives.

Chapter 3

Methodology

3.1 Overview of Methodological Approach

The methodological framework of this research was strategically developed to align with the dual objectives of the study: (1) to evaluate the performance of a sequential electrochemical treatment system combining AEC and PC for the treatment of landfill leachate and (2) to develop and characterize a biomass-derived cathode for enhanced electrochemical advanced oxidation processes. To achieve these objectives, the research methodology was divided into two integrated phases as shown in Fig. 3.1.

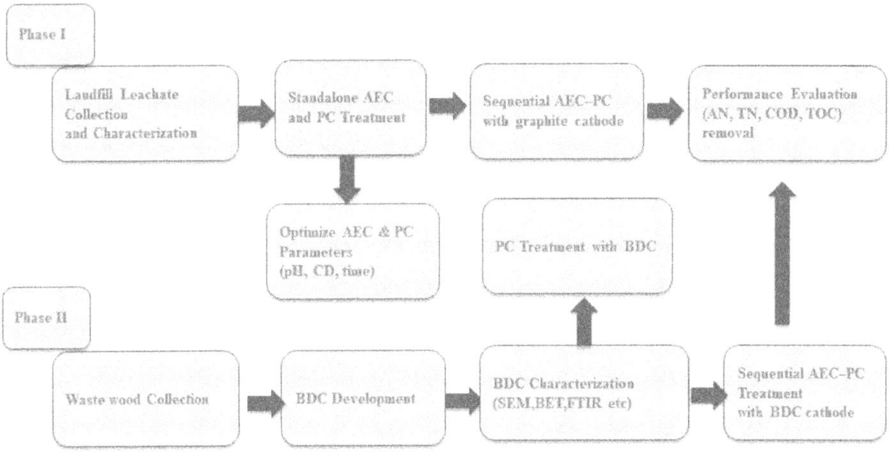

Fig. 3.1: Overall methodology of the research work

The first phase involved optimizing the sequential electrochemical treatment process using conventional electrodes under various operational parameters such as pH, current density, and reaction time. This phase established the fundamental process optimization and synergistic effects of AEC followed by PC on real landfill leachate.

The second phase focused on the development of a sustainable cathode material derived from teak wood biomass. The biochar was prepared using a controlled pyrolysis method and subsequently integrated into the electrochemical setup. Comprehensive material characterization techniques, including surface morphology, crystallinity, elemental composition, electrochemical activity, and thermal stability were employed to understand the properties of the developed cathode.

Throughout both the phases, real landfill leachate was used to ensure process relevancy and application scalability. Pollutant removal efficiencies were assessed based on key indicators such as COD, TOC, AN and TN. The synergistic performance of the sequential AEC-PC process was quantified using a synergistic index, and energy consumption was evaluated to determine the process feasibility.

Table 3.1 Different chemical used for the experimental work

Sl. No	Chemicals / Materials	Make and Purity
1	Ferrous Sulfate	Merck and purity of 99%
2	Potassium Dichromate	Merck and purity of 99.9%
3	Sulphuric acid	Merck and purity of 98%
4	Mercuric Sulphate	Merck and purity of 99%
5	Silver Sulfate	Merck and purity of 99.99%
6	Ammonium iron (II) Sulfate hexahydrate	Merck and purity of 98%
7	Boric acid	Merck and purity of 99.5%
8	Sodium tetra borate	Merck and purity of 99%
9	Sodium hydroxide pallets	Merck and purity of 98%

3.2 Materials

The different materials used in the research work such as various chemicals, electrodes, and biomass materials employed for the synthesis of BDC are discussed in this section.

3.2.1 Chemicals

The chemicals used for the current experimental works are shown in the Table. 3.1 and the chemicals used were procured from Kerala, India. H_2SO_4 (5.0 mM) and NaOH (5.0 mM) were used to adjust pH of the aqueous solution.

3.2.2 Electrodes

The iron plate electrode has been used for the AEC process and was procured from metal industry located in Calicut, Kerala, India. The iron electrode was polished with sand paper (150 mesh) and was washed with deionized water before and after each use to remove the passivation layer. Graphite plate electrode with high hydrogen evolution potential was used as the cathode. Graphite electrodes were purchased from Oriental graphic manufactory, India (apparent density 1.9 g/cm^3, electrical resistivity 12 μohm-m and thermal conductivity 140 Kcal/m-hr-°C). The effective area of all the electrodes used in the electrochemical study was 25 cm^2 (10 cm x 2.5 cm). In the AEC process iron plate has been used as the anode and cathode electrode. Graphite as well as BDC, was used as the cathode and iron electrode has been used as the anode for the peroxicoagulation process.

In the preparation of BDC electrode, the waste wood was collected from local furniture industry and washed for several times to remove undesired compounds and dried prior to the experiments. The waste-wood was initially pre-carbonized in hot air oven followed by complete pyrolysis in a quartz tubular furnace under the inert environment. The detailed preparation of biomass derived electrode is discussed in section 3.8.

3.3 Characterization of Landfill Leachate

The leachate used for the study was collected from a landfill site in Kerala, India. The characterization analysis was performed according to American Public Health Association (APHA) standards. The pH and conductivity of the sample have been measured using Hanna instruments (HI 5522). COD was analyzed by closed reflux titrimetric method. Ammoniacal nitrogen analysis was done by using automatic nitrogen analyzer (KjelTRON-Tulin Equipments). TOC was analyzed using TOC-L Analyzer (Shimadzu). BOD, TDS, TS, TSS and COD of the leachate was analyzed according to APHA standard. The leachate analysis was done thrice to understand the deviation in the analysis due to the dilution of the sample and the standard deviation was calculated and reported accordingly.

3.4 Electrochemical Reactor Configuration and Setup

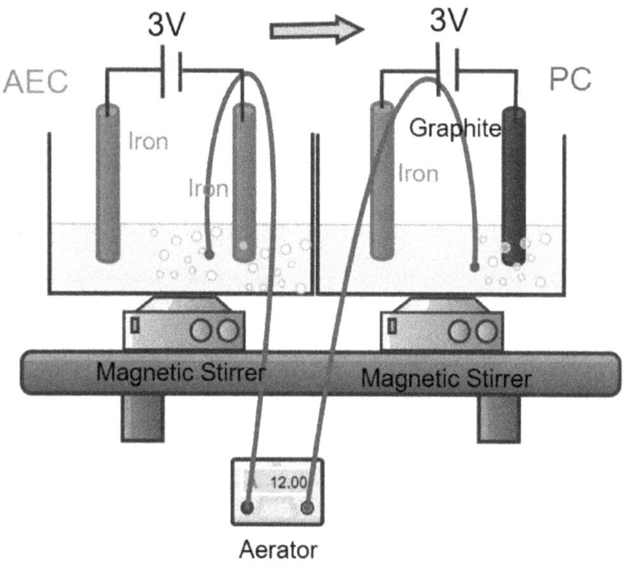

Fig. 3.2: Schematic diagram of the sequential AEC and PC experimental setup (AEC process followed by PC process)

The batch experimental operations have been executed in a 1250 mL electrochemical reactor with a working volume of 1000 mL at room temperature. The electrodes were faced each other and the monopolar electrode configuration was adopted throughout the study. The distance between electrodes were maintained as 25 mm, Direct Current (DC) power supply (GW Instek-GPP-2323) with voltage and current range of 0–30 V and 0–3 A respectively has been used for the electrolytic reaction. The solution was aerated using an aquatic aerator, purchased from the local market. During the AEC process, the aeration was done on the bulk solution and for peroxicoagulation process, aeration was done on the surface of the cathode. For uniform mass transfer a magnetic stirrer (REMI) has been used for bulk solution turbulence and the Rotation Per Minute (RPM) was varied based on the requirements. The schematic diagram of the sequential AEC and PC experimental setup (AEC process followed by PC process) is shown in Fig. 3.2.

3.5 Aerated Electrocoagulation Process and its Optimization

Aerated electrocoagulation is a variant of electrocoagulation in which air (or oxygen-containing gas) is sparged into the electrolytic reactor while a soluble sacrificial anode (typically iron or aluminum) dissolves under an applied current. At the anode, metal ions (e.g., Fe^{2+}/Fe^{3+}) are generated in-situ; in the bulk solution they hydrolyze to form metal hydroxide species that destabilize colloids and complex dissolved organics via charge neutralization, sweep coagulation, and adsorption [153,154,183,184]. Simultaneously, cathodic reactions generate fine hydrogen bubbles that aid flotation, while the externally supplied aeration improves mixing, mass transfer, and DO [183,185]. In this study, AEC was operated under galvanostatic control, with air sparged to the bulk continuously to maintain an oxygenated, well-mixed condition. The study was conducted with a fixed electrode surface area of 25 cm^2 (10 cm x 2.5 cm) and a spacing of 25 mm was maintained throughout the study. The sampling was done for first 120 min of electrolysis at a fixed time interval and the collected samples were analysed for COD, TOC, TN and AN. All the experiments were repeated twice.

3.5.1 Effect of pH

The pH is the single most influential variable in iron-based AEC because it controls the identity and solubility of Fe hydrolysis products, the pollutant speciation / charge, and gas-liquid-solid interactions [142,154,186].

At acidic pH, Fe remains more soluble (Fe^{2+}/Fe^{3+} aquo complexes), favouring charge neutralization but limiting sweep coagulation; coagulant carryover can rise [186-188]. For aerated electrocoagulation, the effects of pH were tested at a constant current density of 30 mA/cm^2 and the pH was varied as 4.5, 7.3 and 10.5 keeping other operating conditions fixed. The pH values were selected based on the literature study. At intervals of 10, 20, 30, 40, 50, 60, 90 and 120 minutes, samples were collected. The collected samples were analysed for COD, TOC, TN and AN.

3.5.2 Effect of Current Density

The current density dictates the rate of all electrochemical events and thus controls both treatment kinetics and resource intensity [189]. For aerated electrocoagulation, the effects of current density were studied at pH 7.3 by varying current density as 15 mA/cm^2, 30 mA/cm^2, and 45 mA/cm^2 keeping other conditions fixed. The current density values were selected based on the preliminary study results.

3.6 Peroxicoagulation Process and Optimization

Peroxicoagulation couples electrocoagulation with Fenton chemistry [110]. In this study, PC process was conducted in batch mode under galvanostatic control. Mechanical mixing of the solution was carried and aerator which ensured uniform reagent distribution and mass transfer was used for aeration. The sampling was done for first 120 min of electrolysis at a fixed time interval and the collected samples were analysed for COD, TOC, TN and AN. All the experiments were repeated twice.

3.6.1 Effect of pH

The pH controls peroxicoagulation kinetics, iron speciation, and H_2O_2 stability thereby shifting the balance between homogeneous oxidation and heterogeneous coagulation [81,190]. In the peroxicoagulation process, Fenton reaction is active in pH range of 2.5–3.5 [191]. For peroxicoagulation, the effects of pH were tested at a constant current density of 30 mA/cm^2, by varying the pH of the aqueous solution as 3, 7.3 and 9. The pH values were selected based on the literature study. At intervals of 15, 30, 45, 60, 90, and 120 minutes, samples were collected. The collected samples were analysed for COD, TOC, TN and AN.

3.6.2 Effect of Current Density

The current density governs the rate at which the PC runs controlling (i) iron dissolution (coagulant + Fenton catalyst) (ii) cathodic processes (H_2/H_2O_2 generation, $Fe^{3+} \rightarrow Fe^{2+}$ regeneration), and (iii) energy use and sludge formation [192]. The influence of current density was studied at pH 7.3 by varying current density as 15 mA/cm^2, 30 mA/cm^2, and 45 mA/cm^2. The current density values were selected based on the literature study.

3.7 Sequential Treatment of Leachate

In the sequential treatment, one treatment is followed by another treatment process. The sequential AEC-PC process is the process where aerated electrocoagulation followed by peroxicoagulation and vice versa in the case of PC-AEC.

3.7.1 Sequential Aerated Electrocoagulation-Peroxicoagulation Process

Initially, AEC treatment has been carried out under the optimal conditions of current density and pH, (Current density 30 mA/cm^2 and pH 7.3) based on the result obtained from the standalone process. The AEC effluent produced under optimal conditions served as the influent for peroxicoagulation process in the sequential AEC-PC treatment. Based on the results observed during the standalone AEC process, the electrolysis time was fixed as 60 min. The pH of the AEC effluent was adjusted to 3 by adding concentrated H_2SO_4 after and the further peroxicoagulation process was facilitated.

Sequential AEC-PC was run for a total of 180 minutes of electrolysis time, which includes 60 min of AEC and 120 min of PC process. At intervals of 0, 10, 20, 30, 40, 50, 60, 75, 90, 105, 120, 150 and 180 minutes, sampling was done and the collected samples were analyzed for COD, TOC, TN and AN. To reduce experimental error, each experiment was run twice at room temperature.

3.7.2 Sequential Peroxicoagulation-Aerated Electrocoagulation Process

Initially PC treatment has been carried out under the optimal conditions (current density of 30 mA/cm^2 and initial pH value of 3. The PC effluent produced under optimal conditions served as the influent for the aerated electrocoagulation process in the sequential PC-AEC treatment. The pH of the PC effluent was adjusted to 7.3 by adding concentrated NaOH after 120 minutes of PC process and further AEC process was facilitated.

Sequential PC-AEC was run for a total of 180 minutes of electrolysis time which includes 120 min of PC and 60 min of AEC process.

3.7.3 Comparison of Sequential Aerated Electrocoagulation-Peroxicoagulation and Peroxicoagulation-Aerated Electrocoagulation Treatment Process

The performance of both the sequential options for the treatment of landfill leachate was compared. The process efficiency of AEC-PC and PC-AEC sequential processes was evaluated based on the TOC, TN, COD and AN of the collected samples with respect to time.

3.7.4 Sludge Quantification of Standalone and Sequential Process

Sludge generation was quantified for the standalone aerated electrocoagulation (AEC, 60 min), peroxicoagulation (PC, 120 min), and the sequential AEC-PC method (180 min). Immediately after each run, the entire treated suspension was transferred from the reactor to a 1 L Imhoff cone (rinsing the reactor walls to recover all solids), allowed to settle under undisturbed condition for a 30 min and the settleable sludge volume was read directly as mL/L. The supernatant was then decanted carefully and the settled sludge was quantitatively transferred to a pre-weighed crucible to determine wet mass of the settled sludge (g/L).

Total solids (TS) and volatile solids (VS) of the sludge were measured by following the standard methods. The dewatered sludge was placed in a pre-weighed crucible, dried at 103–105 °C to constant mass to obtain TS, then ignited at 550 ± 10 °C to obtain VS by loss on ignition; fixed solids (FS) were taken as the difference between TS and VS.

3.8 Preparation of Biomass Derived Cathode

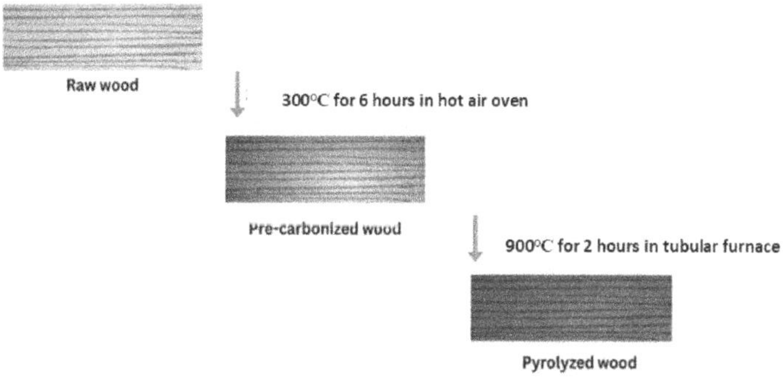

Fig. 3.2 Preparation of wood derived cathode

The waste teak wood was sourced from a furniture industry in Calicut, Kerala. Prior to the study, it was washed multiple times to get rid of unwanted substances and dried. The waste wood was given heat treatment at 300 °C (the temperature was decided as per TGA-DSC results) for 6 hours in hot air oven (forced-convection hot-air type; Binder GmbH, Germany) and undergoes complete pyrolysis (heating of organic material in the absence of oxygen) treatment at 900 °C (fixed as per TGA-DSC results). As shown in the TGA curve (Fig. 5.1(a)), the weight loss of the biomass material stabilizes beyond 500 °C, suggesting the completion of thermal decomposition. Based on this, a pyrolysis temperature of 900 °C was selected to ensure complete carbonization and development of porous carbon structure. The pyrolysis of the wood was done at a heating ramp temperature of 10 °C/min for 2 hours under continuous nitrogen atmosphere in tubular furnace to ensure complete conversion of biopolymers to carbonaceous char, and cooled to ambient temperature under the same inert atmosphere. The wood after pyrolysis is denoted as Biomass derived cathode (BDC) throughout the thesis. The methodology adopted in the preparation of biomass derived cathode is as shown in Fig. 3.2.

3.9 Characterization of Biomass Derived Cathode

A Thermogravimetric (TGA) / Differential scanning calorimetric (DSC) device is used for diagnosing and analyzing processes occurring in a material under different temperature conditions. Differential scanning calorimetric and thermogravimetric (Instruments Q50-0050-0513) analyses of the waste raw wood was conducted using a TA Instruments Q20 & 00200850. The morphology and structure of biomass derived cathode was characterized by using Scanning Electron Microscopy (ZEISS Gemini Sigma 300) coupled with X-ray diffraction (X'Pert3, PANalytical). The X-ray diffraction study was done with the 2θ ranging from 20° to 80°. The surface area of the electrodes was measured by using BET analyzer (Belsorp-max). Raman analysis was used to determine the degree of disorder and the structure of defects in biochar and was assessed by LabRam HR Evolution, Horiba. FTIR graphs are used to determine the chemical functional groups present in the BDC (Perkin Elmer Spectrum Two).

Cyclic voltammetry (CV) was conducted using an electrochemical workstation equipped with a three-electrode system (CH instruments electrochemical analyzer). The working electrodes consisted of BDC and graphite electrode, while the counter electrode was composed of platinum foil. Additionally, Ag/AgCl electrode has served as the reference electrode.

3.10 Peroxicoagulation Process for the Treatment of Leachate Using Biomass Derived Cathode

Electrochemical treatment was carried out using peroxicoagulation process using BDC as cathode and iron electrode as anode. The study was conducted with a fixed electrode surface area of 25 cm^2 (10 cm x 2.5 cm) and a spacing of 25 mm was maintained throughout the study as similar to that of peroxicoagulation process with graphite as cathode. The sampling was done for first 120 min of electrolysis at a fixed time interval and the collected samples were analysed for COD, TOC, TN and AN. All the experiments were repeated twice.

3.10.1 Effect of pH

For peroxicoagulation, the effects of pH were tested at a constant current density of 30 mA/cm^2, by varying the pH of the aqueous solution as 3, and 7.3. The pH values were selected based on the literature study. At intervals of 15, 30, 45, 60, 90 and 120 minutes, samples were collected. The collected samples were analysed for COD, TOC, TN and AN.

3.11 Sequential Treatment of Leachate Using Biomass Derived Cathode

Peroxicoagulation is operated with an iron anode and a BDC cathode (Fe-BDC), exploiting BDC two-electron oxygen reduction to generate H_2O_2 in-situ and to regenerate Fe^{3+} to Fe^{2+} (electro-Fenton synergy). AEC is operated with iron as both anode and cathode (Fe-Fe), emphasizing sweep coagulation and bubble-assisted flotation at near-neutral pH without additional oxidant pathways. Initially PC treatment has been carried out under the optimal conditions (current density of 30 mA/cm^2 and initial pH value of 3). The PC effluent produced under optimal conditions served as the influent for the aerated electrocoagulation process in the sequential PC-AEC treatment. The pH of the PC effluent was adjusted to 7.3 by adding concentrated NaOH after 120 minutes of PC process and further AEC process was facilitated.

Sequential PC-AEC was run for a total of 180 minutes of electrolysis time, which includes 120 min of PC and 60 min of AEC process. At intervals of 0, 15, 30, 45, 60, 90, 120, 130, 140, 150, 160, 170 and 180 minutes, sampling was done and the collected samples were analyzed for COD, TOC, TN and AN.

3.12 Comparison of Sequential treatment of Leachate Using Graphite and Biomass Derived Cathode

The effect of PC process cathode material on the performance of sequential electrochemical treatment was examined. The two different configurations were examined under otherwise identical operating conditions; Fe-Graphite (iron anode with graphite cathode in PC) and Fe-BDC (iron anode with biomass-derived carbon cathode in PC). In all cases, the AEC stage is operated with iron as both anode and cathode (Fe-

Fe) at near-neutral pH. Compared both the treatment aspects in the terms of TOC, TN, COD and AN.

Chapter 4

Sequential Treatment of Leachate with AEC-PC Process

4.1 General

In this study, the performance analysis of electrochemical treatments such as peroxicoagulation process and aerated electrocoagulation process has been separately and sequentially used for the treatment of leachate. Electrodes used for the aerated electrocoagulation process was iron (both anode and cathode) and for the peroxicoagulation process, iron was used as anode and graphite as cathode with an electrode area of 12.5 cm^2. The initial concentration of NH_4–N, Total Nitrogen, COD and TOC of the leachate were 480 mg/L, 997 mg/L, 40200 mg/L and 9850 mg/L respectively. The characteristics of the leachate is shown in Table 4.1. The BOD/COD ratio of the leachate was 0.09 which indicates that the leachate was a stabilized leachate.

4.2 Aerated Electrocoagulation Process

The COD of stabilized landfill leachate is very high, indicating high pollution loads and high concentration of impurities represented by total solids. To reduce the pollution load aerated electrocoagulation has carried out as a pre-treatment.

Electrochemical reactions that occur in solution such as electro dissolution rate, gas evolution, electro flotation, water reactions, etc., as well as their extension was determined by the applied current density [193]. In this study, the different parameters like pH, current density and the electrolysis time has been varied and its influence on the treatment of leachate has been evaluated based on the COD, TOC, NH_4–N and TN removal.

Table 4.1 Characteristics of landfill leachate

Parameter	Concentration in leachate sample (mg/L) except pH and BOD/COD	Standard deviation
pH	7.3	± 0.5
Total Solids	67720	± 2540
Total Dissolved Solids	24060	± 50
Total Suspended Solids	43660	± 1240
COD	40200	± 1050
BOD	3650	± 30
NH_4–N	550	± 25
Total nitrogen	997.5	± 50
Total Organic Carbon	9850	± 100
Chloride	6420.5	± 140
BOD/COD	0.09	

4.2.1 Effect of Current Density

Current density influences the dosing rate of the coagulant particles into the electrolyte's volume, the size and growth of the flocs and the rate at which bubbles are produced, and it is regarded as a crucial factor during electrocoagulation treatment [194]. However, for a higher current density the anodic surface can be isolated by the formation of a stable oxide layers due to the oxidation reactions that promote the corrosion phenomena generating passivation effects.

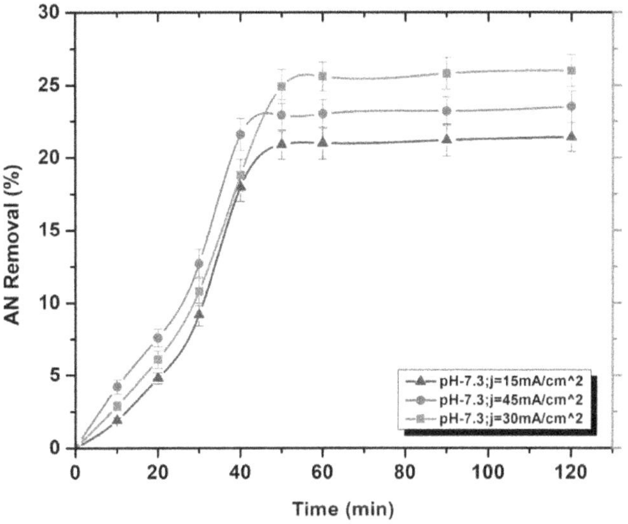
AN Removal (%)
Time (min)
pH-7.3;j=15mA/cm^2
pH-7.3;j=45mA/cm^2
pH-7.3;j=30mA/cm^2

(a)

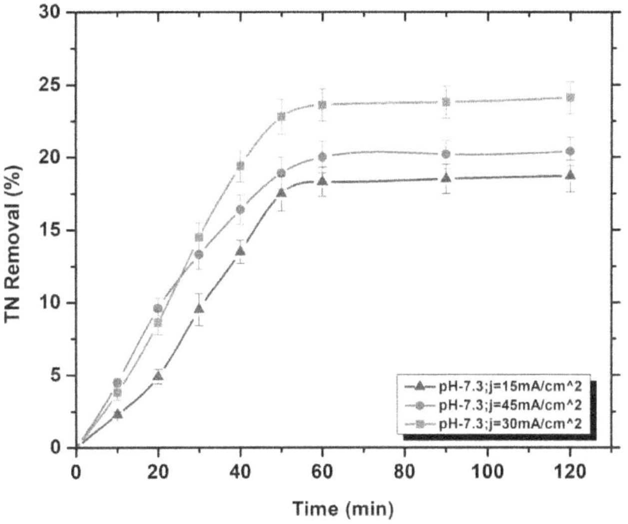
TN Removal (%)
Time (min)
pH-7.3;j=15mA/cm^2
pH-7.3;j=45mA/cm^2
pH-7.3;j=30mA/cm^2

(b)

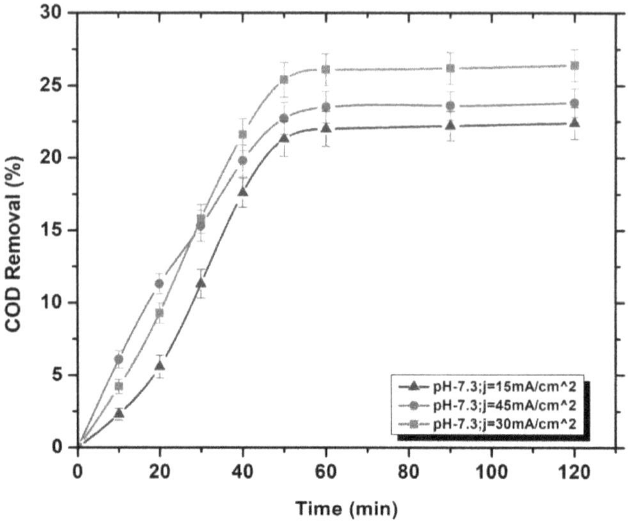

(c)

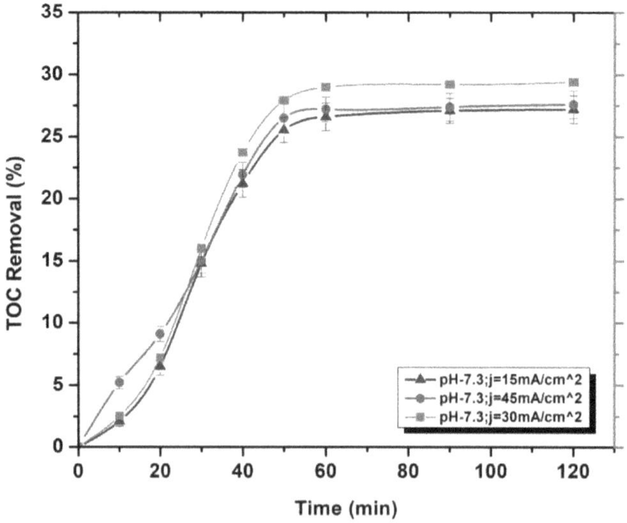

(d)

Fig. 4.1 Effect of Current Density on AEC based on the removal of a) Ammoniacal nitrogen, b) Total Nitrogen, c) COD and d) TOC

Moreover, the higher current density elevates the temperature of bulk solution, and it may result in the increased concentration of stabilized landfill leachate [3]. In this study, the results show that COD, TOC, NH_4–N and TN removal as a function of current density and is shown in Fig. 4.1. Percentage removal of ammoniacal nitrogen was studied for 15 mA/cm^2, 30 mA/cm^2 and 45 mA/cm^2, at a pH of 7.3 (natural pH of Landfill Leachate). Maximum removal efficiency of NH_4–N was 21%, 25.6% and 23% for 15 mA/cm^2, 30 mA/cm^2 and 45 mA/cm^2 respectively for 60 minutes of reaction time shown in Fig. 4.1(a). The removal efficiency of NH_4–N was increased when the current density increased from 15 to 30 mA/cm^2. Similarly, maximum removal of TN at pH 7.3 for 15 mA/cm^2, 30 mA/cm^2 and 45 mA/cm^2 was 18.3%, 23.6.7% and 20% respectively as shown in Fig 4.1(b). When the current density was further increased to 45 mA/cm^2, there is no significant removal because higher current density may result in the formation of a stable oxide layers on the anodic surface that promote the corrosion phenomena and generating passivation effects.

The removal efficiency of COD (Fig. 4.1(c)) has increased when the current density increased from 15 to 30 mA/cm^2. Similarly, 26.6%, 28.9% and 27.2% of TOC (Fig. 4.1(d)) removal has been observed for a current density of 15 mA/cm^2, 30 mA/cm^2 and 45 mA/cm^2 respectively. When the current density was further increased from 45 mA/cm^2, there is no significant removal because higher current density may elevate the temperature of bulk solution, resulting in the increased concentration of stabilized leachate [3]. Increased working current will result in increased energy consumption. During electrocoagulation of chloride-bearing leachate, a portion of the applied charge produces active chlorine (Cl_2/HOCl/ClO^-) at the anode or via mediated reactions in the bulk. These oxidants react with natural organic matter (NOM), amino-N precursors, and halides (Cl^-/Br^-) to form carbonaceous Disinfection By-Products (DBP) (e.g., Trihalomethanes (THM)) and nitrogenous DBPs (e.g., Haloacetonitriles (HAN)). The effect is amplified at higher current density (J) because more charge per unit time is delivered to chlorine-generating reactions, increasing oxidant exposure and DBP yields. In landfill leachate matrices, EC has been shown to generate multiple DBP classes (THMs, HAAs etc), with unregulated DBPs often dominating overall toxicity [195,196]. A current density of 30 mA/cm^2 has given better removal for leachate treatment when compared to 15 mA/cm^2 and 45 mA/cm^2.

4.2.2 Effect of pH

pH is a significant parameter in the electrochemical process [197-199]. It affects the treatment performance because of the formation of metal hydroxides in electrocoagulation (iron hydroxide, Aluminium hydroxide, etc.,) [200,201]. The pH of the solution influences the dissolution of Fe electrode precipitate concentration [28] and thereby influences the formation of flocs. The development of lepidocrocite was favored at pH 7 and the pH range where magnetite and carbonate green rust were best formed was 8 to 9 [202]. At pH greater than 7, various Fe(II, III) species are thermodynamically possible to form [27,28]. In this study weakly alkaline conditions were favouarble for the formation of Fe $(OH)_n$ species and which reduces the pollution loads. To find out the effect of pH, aerated electrocoagulation was performed at pH 4.5 and 10.5, in addition to the pH 7.3. The maximum removal efficiency of NH_4–N at pH 4.5 and 10.5 were 15.6% and 17.1% respectively at 30 mA/cm^2 shown in Fig. 4.2(a). Similar trend was followed in the case of TN as shown in Fig. 4.2(b).

The maximum removal efficiency of NH_4–N at pH 4.5 and 10.5 were 15.6% and 17.1% respectively at 30 mA/cm^2. However, at natural pH 7.3 removal efficiency was 26.7% for NH_4–N because weakly alkaline conditions were favourable for the formation of $Fe(OH)_n$ species and which reduces the pollution loads. When increase in pH (> 4.5) of leachate sample, consequently there is an increase in the concentration of dissolved iron and thereby increased formation of iron hydroxide which degrade the ammoniacal nitrogen in the electrochemical process [203].

More over the degradation of ammoniacal nitrogen is significantly depend on residual chlorine. Predominated chlorine species for a pH range of 1-3 is chlorine. But hypochlorous acid predominates in 5-7 pH range. Above pH > 8 ClO^- prevails as the main species. HOCl is more efficient than the other two chlorine species for removing NH_4–N due to its high oxidation potential (1.63 V) [204]. The formation of these species in water medium as in Eq. (4.1-4.3) [116].

$$2Cl^- \rightarrow Cl_2 + 2e^- \quad (4.1)$$

$$Cl_2 + H_2O \rightarrow HOCl + H^+ + Cl^- \quad (4.2)$$

$$HOCl \rightarrow H^+ + OCl^- \quad (4.3)$$

Ammonia nitrogen removal effectiveness in landfill leachate is significantly influenced by the chloride ion content. Chlorinated derivatives of ammonia are created as a result of the interaction between Eqs. (4.4) and (4.5). Chlorimide formed will hydrolyze these molecules and the ammonia nitrogen will get oxidised as per Eqs. (4.6) and (4.7) [205].

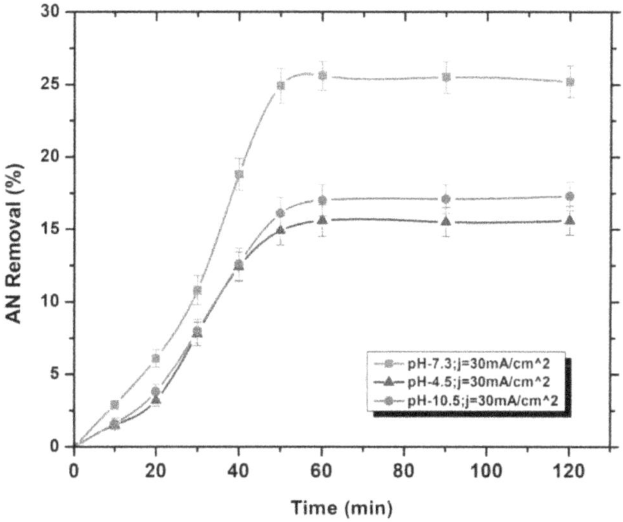

(a)

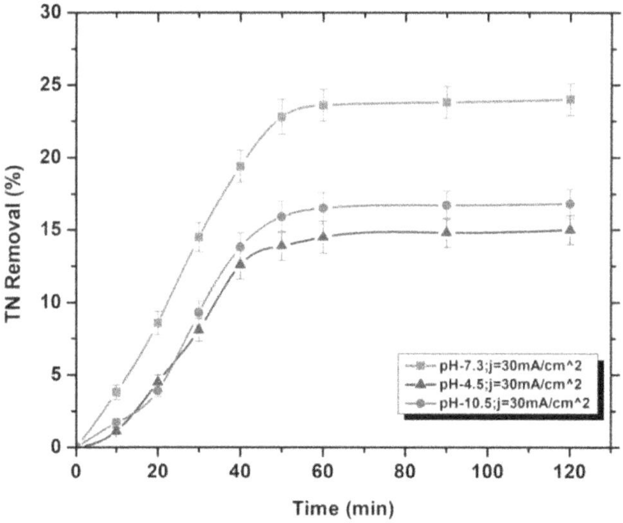

(b)

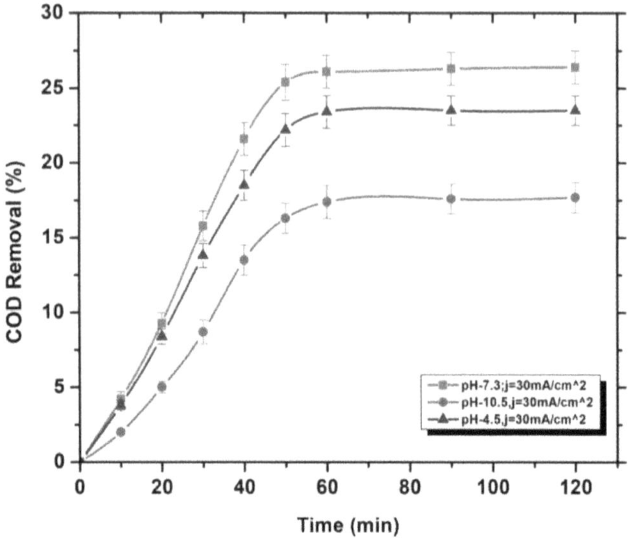

(c)

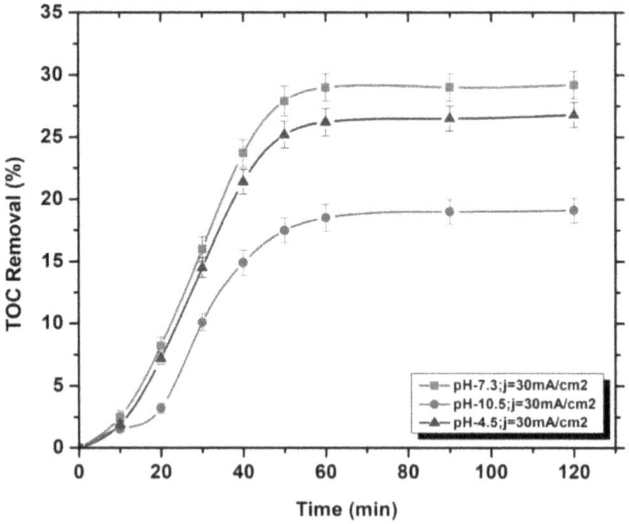

(d)

Fig. 4.2 Effect of pH on the AEC based on removal of a) Ammoniacal nitrogen, b) Total Nitrogen, c) COD and d) TOC

$$NH_4^+ + HClO_{(aq)} \rightarrow NH_2Cl + H_2O + H^+ \quad (4.4)$$

$$HClO_{(aq)} + NH_2Cl \rightarrow NHCl_2 + H_2O \quad (4.5)$$

$$NHCl_2 + H_2O \rightarrow NOH + 2H^+ + 2Cl^- \quad (4.6)$$

$$NHCl_2 + NOH \rightarrow N_2 + HClO + H^+ + Cl^- \quad (4.7)$$

The maximum removal efficiency of COD at pH 4.5 and 10.5 were 23.4% and 17.4% respectively as shown in Fig. 4.2(c). At sub-acidic pH 4.5, iron remains predominantly as $Fe^{2+}{}_{(aq)}$ (and Fe^{3+} is strongly hydrolyzed-suppressed), so the formation of Fe(III) hydroxide solids the main 'sweep-floc' phase is limited within the same residence time. Near-neutral pH (7.3) favors rapid Fe^{2+} to Fe^{3+}conversion and hydrolysis, yielding abundant $Fe(OH)_3(s)$ and stronger sweep-floc coagulation and in higher pH 10.5 COD removal is decreased due to increased concentration of charged ferric complexes which will remove the pollutants selectively accordance with charge [116] and for TOC the maximum removal efficiency was 26.2% and 18.5% for the pH 4.5 and 10.5 respectively as shown in Fig. 4.2(d).

From the results (Fig. 4.2(c)) it is observed that a maximum of 26.2% COD removal efficiency has been achieved at 60 min. The concentration of iron ions and associated hydroxide flocs, as well as the rate of bubble production, rise as the electrolysis time increases. Coagulation and flotation had the effect of removing the contaminants from the leachate [206]. After 60 min, removal rate has not increased to a considerable range, which might be due to passivation layer on the electrode material [207] because the solution's concentration of OH^- increased with increased working time, there was a larger capacity for oxygen to react with the iron on the electrodes surface and form the metal oxide coating. This metal oxide coating, sometimes referred to as a passive oxide film is formed. The removal of pollutants reduced as a result of the thicker metal oxide film that formed on the electrode plates when the gas bubbles were created in large quantities [208]. Moreover, the longer the electrolysis time, the higher the floc concentration in the electrochemical reactor, leading the increase in the pH [209].

4.3 Peroxicoagulation process

All electrolytically produced ferrous ions undergo Fenton's reactions below pH 4, which is the primary cause of the predominate degradation mechanism of the peroxicoagulation process. In the peroxicoagulation process, ferrous ions (Eq. (4.8)) and hydrogen peroxide are produced (Eq. (4.9)) in the cell during the electrolysis. The organic pollutants are removed by degradation and coagulation i.e.; it's a combination of the electro-Fenton and aerated electrocoagulation processes. In this study, the different parameters like pH, current density and the electrolysis time has been varied and its influence on the treatment of leachate has been evaluated based on the COD, TOC, NH_4–N and TN removal.

4.3.1 Effect of Current Density

In the electrochemical process, the applied current density is regarded as the electrokinetic regulating factor [117,210]. Current density controls how much $HO^{\bullet}$ radicals are produced (Eq. (4.10)), how much H_2O_2 is cathodically added to the solution, and how quickly the reduction of Fe^{3+} to Fe^{2+} occurs [117,211]. Initially, the effect of current density was observed for peroxicoagulation process by varying the current density from 15 to 45 mA/cm^2 by keeping the pH at 7.3, and the results are as shown in Fig. 4.3.

The results shows that COD, TOC, NH_4–N and TN removal were influenced by current density. The removal of NH_4–N with respect to current density is shown in Fig. 4.3(a). Maximum removal efficiency of NH_4–N was observed as 30.8%, 34% and 32.3% for 15 mA/cm^2, 30 mA/cm^2 and 45 mA/cm^2 respectively for 120 min of reaction time. When the current density was increased from 15 mA/cm^2 to 30 mA/cm^2 the removal of NH_4–N was increased from 30.8% to 34%, the increase in current density accelerates the formation of larger hydroxyl radicals and $Fe(OH)_3$ flocs [212]. In addition, higher current densities accelerate the rate of active chlorine production, which aid in the removal of NH_4–N. Whereas, when current density increased from 30 mA/cm^2 to 45 mA/cm^2 the removal of of NH_4–N was declined from 34% to 32.3% it may be due to the fact that the increased current density elevates temperature of bulk solution resulting increased concentration of stabilized landfill leachate. Moreover, side parasitic reactions occur which consume $HO^{\bullet}$ radicals [213]. At higher current density there are more Fe^{2+} and H_2O_2 ions in the bulk solution, and that the excess Fe^{2+} ions react with $HO^{\bullet}$ radicals to produce excess Fe^{3+} Eq.

(4.11). These ferric ions interact with H_2O_2 to produce hydroperoxyl radicals, which are weaker oxidant than $HO^\bullet$ as per Eq. (4.12). This radical is also produced by the reaction between the H_2O_2 and $HO^\bullet$ as per Eq. (4.13).

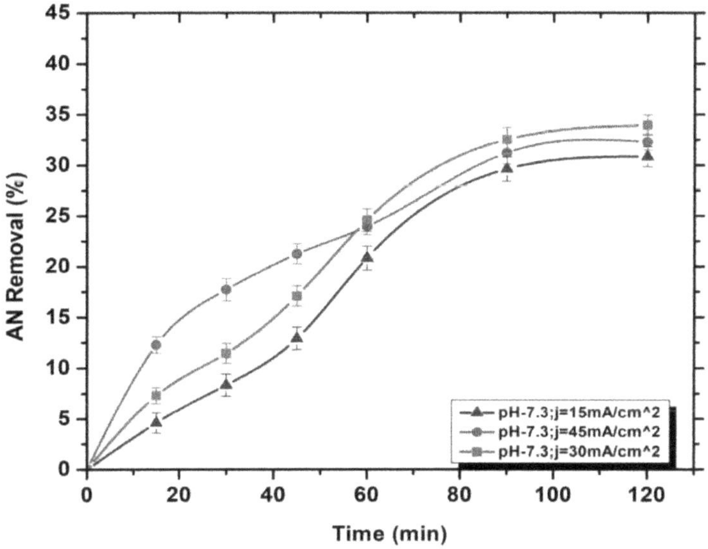

(a)

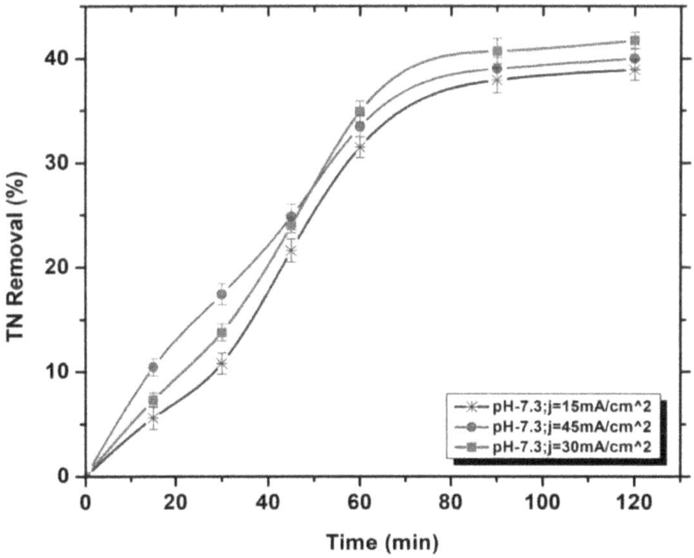

(b)

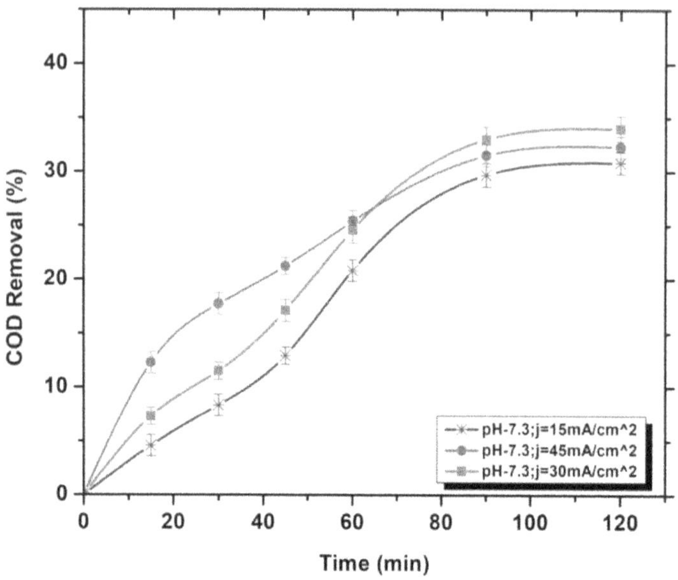

(c)

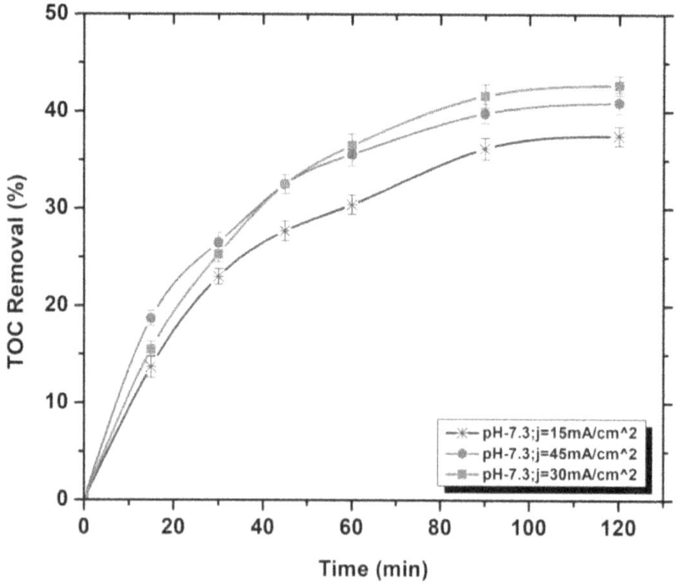

(d)

Fig. 4.3 Effect of Current Density on PC based on removal of a) Ammoniacal nitrogen, b) Total Nitrogen, c) COD and d) TOC

In addition, the dimerization of $HO^{\bullet}$ into H_2O_2 is a possibility Eq. (4.14) [110,210, 213-215]. Also, the high-power consumption caused by side reactions including heat generation and oxygen evolution at high current densities, the NH_4–N removal rate did not increase linearly with the rising current density [216].

$$2Fe \rightarrow 2Fe^{2+} + 4e^- \quad (4.8)$$

$$O_2 + 2H^+ + 2e^- \rightarrow H_2O_2 \quad (4.9)$$

$$Fe^{2+} + H_2O_2 \rightarrow Fe^{3+} + OH^- + HO^{\bullet} \quad (4.10)$$

$$Fe^{2+} + HO^{\bullet} \rightarrow Fe^{3+} + OH^- \quad (4.11)$$

$$Fe^{3+} + H_2O_2 \rightarrow Fe^{2+} + H^+ + HO_2^{\bullet} \quad (4.12)$$

$$H_2O_2 + HO^{\bullet} \rightarrow HO_2^{\bullet} + H_2O \quad (4.13)$$

$$2HO^{\bullet} \rightarrow H_2O_2 \quad (4.14)$$

The $HO^{\bullet}$ formed via Eq. (4.10) reacts with NH_4^+–N/NH_3 –N to give the following products and is represented in Eq. (4.15) [217]. Cathode reduction was the primary reaction that removed nitrogen in the presence of an iron anode Eq. (4.16). In general, nitrogen gas was the primary by-product of nitrite-N reduction [218,219]. The removal of TN with respect to current density is shown in Fig. 4.3(b). The maximum removal efficiency of TN observed was 38.9%, 41.7% and 40% for 15 mA/cm^2, 30 mA/cm^2 and 45 mA/cm^2 respectively for 120 min of reaction time. When the current density was increased from 15 mA/cm^2 to 30 mA/cm^2 the removal of TN got increased from 38.9% to 41.7%. When the current density was further increased from 30 mA/cm^2 to 45 mA/cm^2, the removal of TN was declined from 41.7% to 40% it may be due to fact that the high current density will encourage competition between the NO_x–N reduction, ORR and hydrogen evolution reaction on the graphite cathode. This competition reaction is not favorable for the in-situ electro generation of H_2O_2, leading to a drop in the removal efficiency [3].

$$NH_4^+ - N + HO^{\bullet} \rightarrow NH_2OH \rightarrow NOH \rightarrow NO \rightarrow HNO_2 \rightarrow HNO_3 \quad (4.15)$$

$$2NO_2^- + 8H^+ + 6e^- \rightarrow N_2 + 4H_2O \quad (4.16)$$

The removal of COD with respect to current density is shown in Fig. 4.3(c). Maximum removal efficiency of COD observed was 31%, 34% and 32% for 15 mA/cm^2, 30 mA/cm^2 and 45 mA/cm^2 respectively for 120 min of reaction time. When the applied current density was increased from 15 mA/cm^2 to 30 mA/cm^2 the removal of COD was increased from 31% to 34%, it's due to the fact that increased H_2O_2 production on the graphite cathode and the increased formation of $Fe(OH)_3$ precipitates at higher current levels [220]. When the current density was further increased from 30 mA/cm^2 to 45 mA/cm^2, the removal of COD was declined from 34% to 32%. It may be due to the fact that undesired parasitic processes like H_2O_2 decomposition may occur Eq. (4.17) [221,222].

$$2H_2O_2 \rightarrow 2H_2O + O_2 \uparrow \tag{4.17}$$

The removal of TOC with respect to current density is shown in Fig. 4.3(d). Maximum removal efficiency of TOC observed was 37.5%, 42.7% and 40.9% for 15 mA/cm^2, 30 mA/cm^2 and 45 mA/cm^2 respectively for 120 min of reaction time. When the applied current density was increased from 15 mA/cm^2 to 30 mA/cm^2 the removal of COD was increased from 37.5% to 42.7% due to the fact that increased H_2O_2 production on the graphite cathode and the increased formation of $Fe(OH)_3$ precipitates at higher current levels [220]. The removal of NH_4–N competed with the removal of organic matter in a combined removal process [219,223]. During the initial 25 min of electrolysis time at 30 mA/cm^2, concentration of TOC showed a significant reduction due to reduction in concentration of ammoniacal nitrogen. When the current density was further increased from 30 mA/cm^2 to 45 mA/cm^2, the removal of COD was declined from 42.7% to 40.9%. It may be due to the fact that side reactions occur that can consume the highly reactive $HO^{\bullet}$ and thus decrease the oxidizing efficiency as per Eq. (4.18) [221,222].

$$H_2O_2 + HO^{\bullet} \rightarrow HO_2^{\bullet} + H_2O \tag{4.18}$$

4.3.2 Effect of pH

The pH is an important parameter which influences the PC. To investigate the effect of pH, peroxicoagulation was carried out at pH 3 and 9, in addition to 7.3. The removal of COD with respect to pH is shown in Fig. 4.4(c). The maximum removal efficiency of COD observed at pH 3, 7.3 and 9 were 37.3% and 34% and 32% respectively at 30 mA/cm^2. When the pH is dropped from 7.3 to 3 the removal is increased from 34% to

37.3%, since removal of COD occurs at pH greater than 3.5 is due to coagulation and sorption on iron hydroxide complexes [224]. Moreover, the rapid degradation happens at pH 3, since there is large electrochemical production of H_2O_2 [224,225]. Due to the stability of H_2O_2, which begins to rapidly breakdown into molecular oxygen without creating an $HO^{\bullet}$ at pH levels greater than 3, oxidation efficiency declines [224,226].

In addition to formation of $HO^{\bullet}$ via the reaction of H_2O_2 and Fe^{2+} Eqs. (4.19) and (4.20) [218] and the hydroxyl radical reacts with organic species as given in Eq. (4.21). Apart from the $HO^{\bullet}$ various reactive chlorine species also formed during the electrolysis with the presence of Cl^- ion. The reactive chlorine species includes chlorine monoxide radical ($ClO^{\bullet}$), hydroxidochlorate radical ($HOCl^{\bullet-}$), chlorine radical ($Cl^{\bullet}$), and dichloride radical ($Cl_2^{\bullet-}$) as shown in Eqs. (4.22-4.28). Such reactive chlorine species also significantly aid in the breakdown of organic pollutants, particularly those that contain functional groups rich in electrons [227]. The COD removal efficiency from Fig. 4.4(c) shows that as the reaction time increased COD removal in the peroxicoagulation process improved. During the first 90 min of treatment the wastewater was rapidly degraded as largely due to the easily degradable organics.

$$O_2 + 2H^+ + 2e^- \rightarrow H_2O_2 \quad (4.19)$$

$$Fe^{2+} + H_2O_2 \rightarrow Fe^{3+} + OH^- + HO^{\bullet} \quad (4.20)$$

$$RH + HO^{\bullet} \rightarrow R^{\bullet} + OH \quad (4.21)$$

$$HOCl + HO^{\bullet} \rightarrow ClO^{\bullet} + H_2O \quad (4.22)$$

$$OCl + HO^{\bullet} \rightarrow ClO^{\bullet} + OH^- \quad (4.23)$$

$$Cl^- + HO^{\bullet} \rightleftarrows HOCl^{\bullet-} \quad (4.24)$$

$$HOCl^{\bullet-} + H^+ \rightleftarrows Cl^{\bullet} + H_2O \quad (4.25)$$

$$HOCl + Cl^{\bullet} \rightarrow H^+ + Cl^- + ClO^{\bullet} \quad (4.26)$$

$$ClO^- + Cl^{\bullet} \rightarrow Cl^- + ClO^{\bullet} \quad (4.27)$$

$$Cl^- + Cl^{\bullet} \rightarrow Cl_2^{\bullet-} \quad (4.28)$$

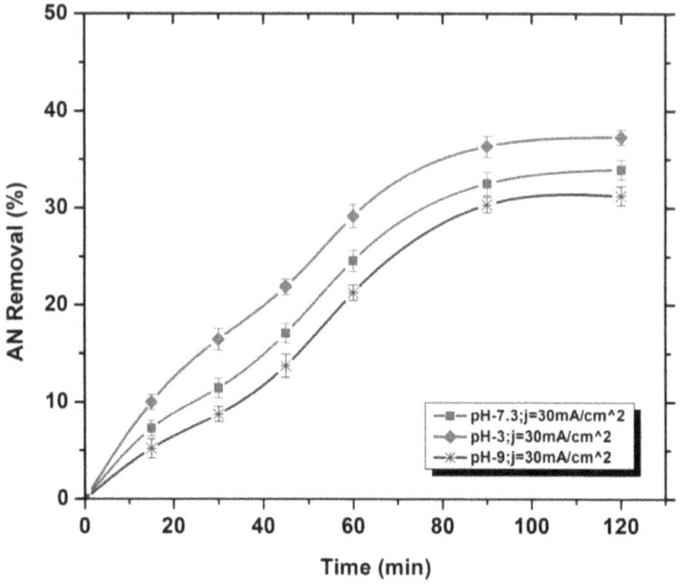
50
40
30
20
10
0
AN Removal (%)
0
20
40
60
80
100
120
Time (min)
pH-7.3;j=30mA/cm^2
pH-3;j=30mA/cm^2
pH-9;j=30mA/cm^2

(a)

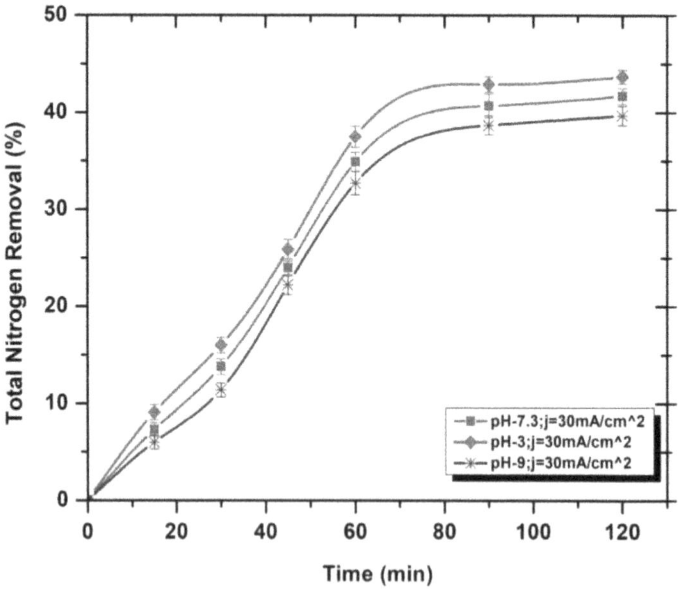
50
40
30
20
10
0
Total Nitrogen Removal (%)
0
20
40
60
80
100
120
Time (min)
pH-7.3;j=30mA/cm^2
pH-3;j=30mA/cm^2
pH-9;j=30mA/cm^2

(b)

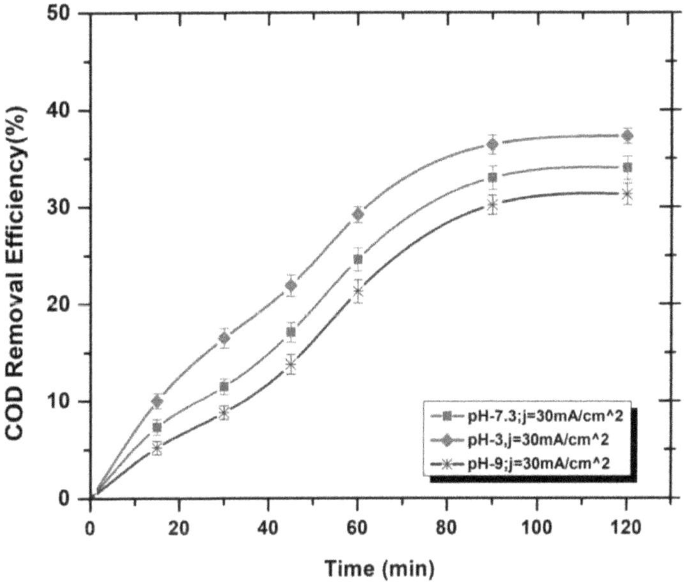

(c)

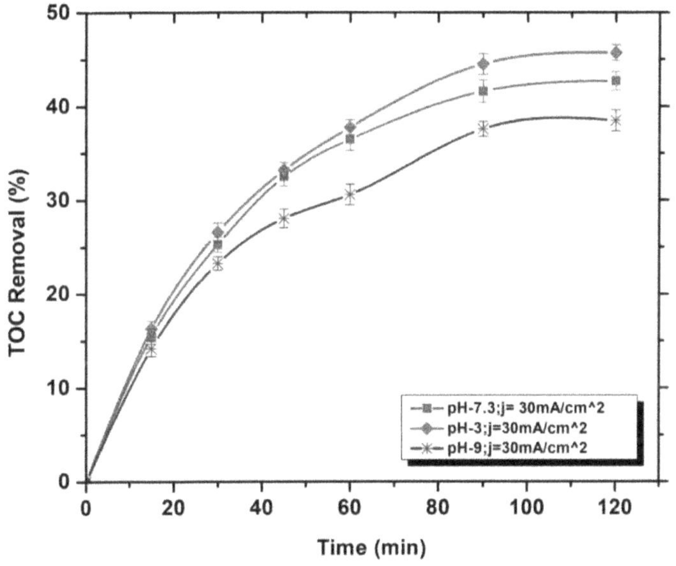

(d)

Fig. 4.4 Effect of pH on PC process based on removal of a) Ammoniacal nitrogen, b) Total Nitrogen, c) COD and d) TOC

After the first 90 min, the degradation then gradually slowed down, following a steady trend through the peroxicoagulation treatment [70, 228].

When pH is increased to 9 the removal was declined from 34% to 32%. It may be due to the instability of H_2O_2, which leads to rapid breakdown into molecular oxygen without creating an $HO^{\bullet}$ at pH > 3, hence oxidation efficiency declines and also in alkaline condition formation of ferric hydroxide complexes leads to a reduction in $HO^{\bullet}$ [224,226]. When pH is increased from 3 the dominating pollutant removal mechanism is electrocoagulation in the PC process rather than electrooxidation [30].

The removal of TOC with respect to pH is shown in Fig. 4.4(d). The maximum removal efficiency for TOC observed at pH 3, 7.3 and 9 were 45.7%, 42.7% and 38.5% respectively at 30 mA/cm^2. When pH is dropped from 7.3 to 3 the TOC removal is increased from 42.7% to 45.7%, it may be due to the fact that precipitation of humic acid compounds in acidic circumstances, the TOC content declines when the initial pH was adjusted to 3.0 without electrochemical treatment [222] and further removal happened after the electrochemical treatment. Moreover, mineralization of organics present in leachate takes place in the pH range of 3-3.5 because pH < 3.5, the solubility equilibrium of iron is controlled by Fe^{3+} ions than other iron hydroxide complexes [30].

The removal of NH_4–N with respect to pH is shown in Fig. 4.4(a). The maximum removal efficiency of NH_4–N observed at pH 3,7.3 and 9 were 37.3%, 34% and 31.2% respectively at 30 mA/cm^2. When pH is dropped from 7.3 to 3 the NH_4–N removal is increased from 34% to 37.3%, may be due to the fact that under neutral or alkaline conditions Fe^{2+} activated H_2O_2 was reduced, and H_2O_2 was decomposed into O_2 and H_2O in alkaline solution [3,191]. The $HO^{\bullet}$ formed due to the Eqs. (4.19) and (4.20) reacts with the NH_4^+–N/NH_3–N to give the products as represented in Eq. (4.21). The Mulliken population analysis revealed that $HO^{\bullet}$ and NH_4–N had electron-rich groups that were comparable, which supported the ineffective removal of NH_4–N by $HO^{\bullet}$. The generation of $ClO^{\bullet}$ accelerated the oxidation of NH_4–N which is formed by the interaction between $HO^{\bullet}$ and active chlorine as given in Eqs. (4.22) and (4.23) and it's reaction rate with the NH_4–N is 17 fold higher than that of $HO^{\bullet}$ [229]. According to Eq. (4.29) direct oxidation of NH_4–N takes place and its indirect oxidation takes place in the presence of Cl^-as per Eqs. (4.30) and (4.31) [230,231]. Apart from this reactions N-centered radicals ($HN_2^{\bullet}$,

$NH_2O^{\bullet}$ and $N^{\bullet}HCl$) would be formed due to the H abstraction by $ClO^{\bullet}$ and $HO^{\bullet}$, and these N-centered radicals are unstable and can be further converted into N_2 as given in Eqs. (4.32)-(4.43) [198,229,232–234]. When pH is increased to 9 the removal was declined from 34% to 31.25% it may be due to higher pH levels hinder the production of $HO^{\bullet}$ because H_2O_2 decomposes into O_2 gas and loses its capacity to oxidise [217].

The removal of TN with respect to pH is shown in Fig. 4.4(b). The maximum removal efficiency of TN observed at pH 3,7.3 and 9 were 43.7%, 41.7% and 39.7% respectively at 30 mA/cm^2. When pH is dropped from 7.3 to 3 the TN removal is increased from 41.7% to 43.7%. Apart from the reduction in NH_4–N concentration there is reduction of NO_x–N to nitrogen due to cathodic reduction thereby significant reduction in TN concentration Eq. (4.16) at pH 3 [235,236]. When pH is increased to 9 the removal was declined from 41.7% to 39.7% it may be due to the fact that the high pH led to a high conversion tendency of NO_x –N to $NH_4{}^+$–N, which was not conducive to TN removal [235].

$$2NH_4^+ + 3OH^- \rightarrow N_2 + 6H_2O + 2H^+ + 6e^- \tag{4.29}$$

$$2NH_4^+ + 3HOCl \rightarrow N_2 + 3H_2O + 5H^+ + 3Cl^- \tag{4.30}$$

$$2NH_4^+ + 3ClO^- \rightarrow N_2 + 3H_2O + 2H^+ + 3Cl^- \tag{4.31}$$

$$NH_3 + HO^{\bullet} \rightarrow N^{\bullet}H_2 + H_2O \tag{4.32}$$

$$HN_2^{\bullet} + HO^{\bullet} \rightarrow NH_2OH \tag{4.33}$$

$$NH_2OH + HO^{\bullet} \rightarrow NH_2O^{\bullet} \rightarrow N_2 \tag{4.34}$$

$$NH_2OH + HO^{\bullet} \rightarrow NO_2^- \rightarrow NO_3^- \tag{4.35}$$

$$NH_3 + ClO^{\bullet} \rightarrow N^{\bullet}H_2 + HOCl \tag{4.36}$$

$$N^{\bullet}H_2 + HOCl \rightarrow NH_2Cl + HO^{\bullet} \tag{4.37}$$

$$NH_2Cl + ClO^{\bullet} \rightarrow N^{\bullet}HCl + HOCl \tag{4.38}$$

$$N^{\bullet}HCl + HOCl \rightarrow NHCl_2 + HO^{\bullet} \tag{4.39}$$

$$NHCl_2 + H_2O \rightarrow NOH + 2HCl \tag{4.40}$$

$$NH_2Cl + NOH \rightarrow N_2 + H_2O + H^+ + Cl^- \tag{4.41}$$

$$NHCl_2 + NOH \rightarrow N_2 + HOCl + H^+ + Cl^- \quad (4.42)$$

$$NHCl_2 + NH_2Cl \rightarrow N_2 + 3H^+ + 3Cl^- \quad (4.43)$$

4.4 Sequential Treatment of Leachate

In this section the sequential treatment of landfill leachate has been discussed in detail. In the sequential treatment AEC followed by PC and PC followed by AEC options were discussed and the results were compared. Based on the AEC and PC process the optimum conditions has been fixed, accordingly the electrolysis time for AEC was fixed as 60 min and PC as 120 min.

4.4.1 Sequential Aerated Electrocoagulation-Peroxicoagulation Process

The combined effect of AEC and PC was studied at the optimal conditions observed during the treatment of leachate in the stand-alone process. The removal of NH_4–N, TN, COD and TOC were studied and the results are shown in Fig. 4.5. AEC was carried out for initial 60 min at pH 7.3, and a current density of 30 mA/cm^2 with iron as both anode and cathode, then the peroxicoagulation was continued for another 120 min at pH 3 and current density of 30 mA/cm^2 with iron as anode and graphite as cathode. Aerated electrocoagulation was performed as a pretreatment, before performing PC.

After 60 min, peroxicoagulation was initiated and the influent had a concentration of 405.3 mg/L, 760 mg/L, 27,697 mg/L and 6993.5 mg/L of ammoniacal nitrogen, total nitrogen, COD and TOC respectively. The maximum removal achieved was 63%, 68%, 78% and 75% for ammoniacal nitrogen, total nitrogen, COD and TOC respectively at the end of electrolysis. The percentage removal of COD progressively increased from 10 min to 50 min. After 50 min of time, lesser effect on COD removal percentage was observed until the next process was started. COD removal was 17.5% during the first 30 min of AEC and it reached to 28% within next 20 min which can be attributed to the fact that metal (oxy) hydroxide flocs produced which neutralize the pollutants which favor the aggregation and settling of suspended particles and dissolved organic matter resulted in COD reduction. In the next process i.e., peroxicoagualtion process the COD removal was 65% on initial 30 min and it reached to 78% in next 90 min which can be attributed to the fact that at initial 30 min the dominant pollutant removal mechanism was electro-Fenton and as the electrolysis time increases there is increase in solution pH which leads to domination of electrocoagulation in the later stages of treatment [30].

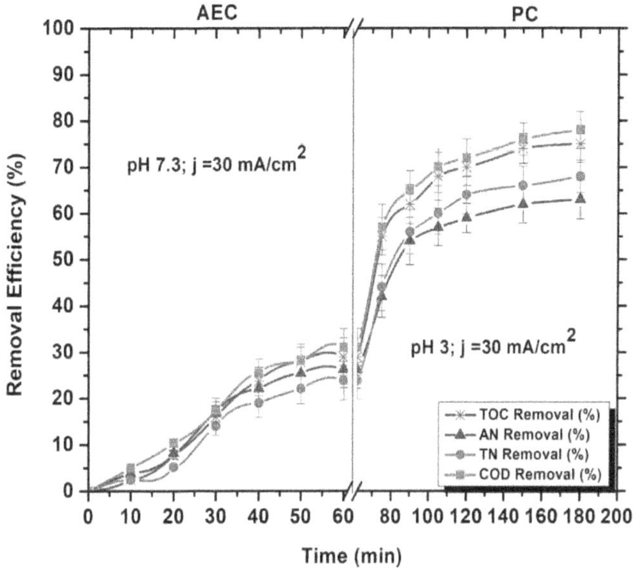

Fig. 4.5. Sequential AEC – PC process for the removal of NH_4-N, TN, COD and TOC (AEC: pH 7.3, Current density 30 mA/cm^2, electrolysis time of 60 min. PC: pH 3, Current density 30 mA/cm^2, electrolysis time of 120 minutes).

The removal efficiency increases in the first 50 min of PC process and further with increase in electrolysis, the removal efficiency shows a lower rate of increase. When the total electrolysis time reaches 180 min, the increase in removal efficiency has become almost saturated. Due to the longer electrolysis period (180 min), there were more soluble Fe ions and this leads to electrode poisoning, which causes the anode's catalytic activity to decline and a polymer coating or film to form on the anode surface. Fig. 4.6 shows the comparison of COD removal with standalone AEC and PC with sequential AEC-PC process. The induction of water on that surface, leads to reaction with active oxygen or an inorganic oxidant, like chloride ions where oxidation may be controlled [237]. Additionally, when the electrolysis time climbed, the rate of bubble production increased and iron hydroxide flocs were generated. That may be the reason for the stable removal of pollutants. It is clear that after 180 min of reaction time the increment in the increase of removal efficiency decreased and the attained a saturation state of removal further increase in the current density also cannot initiate much removal of pollutants from the landfill leachate.

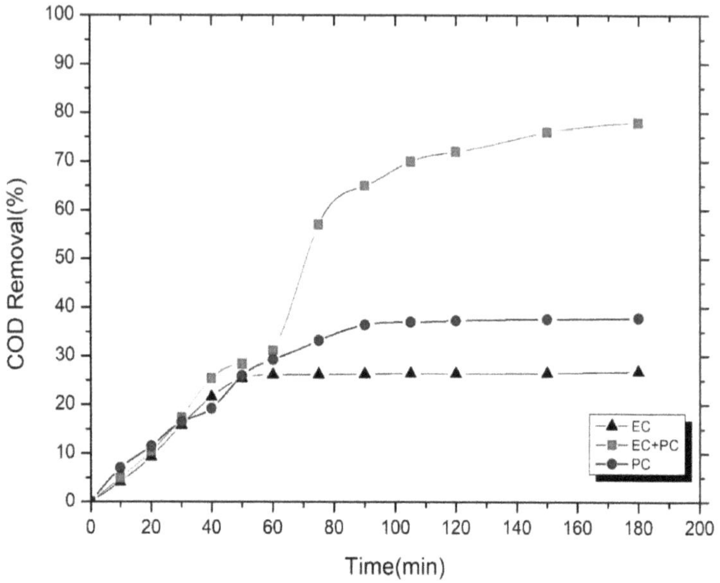

Fig. 4.6. Comparison of COD removal sequential AEC-PC process with standalone AEC and PC process

AEC process followed by degradation process is effective for high- strength compared to other sequential processes, since degradation process cannot carry effectively. Also, the electrochemical advanced oxidation process is unsuitable for treating high-strength wastewater due to insufficient hydrogen peroxide production and excessive electrode passivation. Initial treatment with AEC reduces the pollutant content in wastewater significantly. The AEC greatly reduces the pollutant amounts in wastewater in these configurations (AEC+EAOP), aiding subsequent degradation processes [238,239].

The Fig. 4.6 shows the comparison of COD removal with standalone aerated electrocoagulation process, peroxicoagulation process and with sequential aerated electrocoagulation and peroxicoagulation process under optimum conditions. The sequential process is more effective with a removal efficiency of 78% compared with the standalone aerated electrocoagulation process with 26.2% COD removal under the optimum conditions of pH 7.3, current density 30 mA/cm^2 and peroxicoagulation process with 37.3% COD removal under the optimum conditions of pH 3, current density 30 mA/cm^2.

4.4.2 Comparison of Sequential Aerated Electrocoagulation-Peroxicoagulation and Peroxicoagulation-Aerated Electrocoagulation Treatment Process

Peroxicoagulation process followed by aerated electrocoagulation (PC-AEC) and aerated electrocoagulation followed by peroxicoagulation (AEC-PC) under the identical total treatment condition has been compared for the removal of TOC and the results are shown in Fig. 4.7.

The PC process has been operated for an electrolysis time of 120 min, at a current density of 30 mA/cm^2, pH of 3 and AEC process has been operated for an electrolysis time of 60 min, current density of 30 mA/cm^2, pH of 7.3. The same operation condition has been adopted for both the sequential process PC-AEC and AEC-PC. In AEC, both electrodes used were iron and in PC, the iron electrode was used as anode and cathode used was graphite. When, the PC process was adopted first (PC-AEC), TOC removal increased gradually as shown in Fig. 4.7(a), during the acidic PC stage, reaching 45% by 120 min, subsequent AEC step delivered only a modest incremental gain, ending near 65% at 180 min. This pattern indicates that initiating with oxidation in a matrix rich in radical scavengers (humics, bicarbonate, suspended solids) diverts a significant fraction of $HO^{\bullet}$ away from mineralization. Although partial breakdown to smaller, more coagulable fragments occurs, the later AEC step functions mainly as a turbidity-polishing operation with sweep flocculation removing the particulates efficiently but contributes comparatively little to further TOC abatement.

In contrast as shown in Fig. 4.7(b), when AEC preceded PC (AEC-PC), the first 60 min at pH 7.3 yielded a modest TOC removal of 25% as sweep flocculation debulked colloids and hydrophobic macromolecules. Upon switching to PC and acidifying to pH 3, TOC removal jumped sharply to 55% within further 15 min of electrolysis. The removal of TOC then continued to rise, approaching 75% by 180 min.

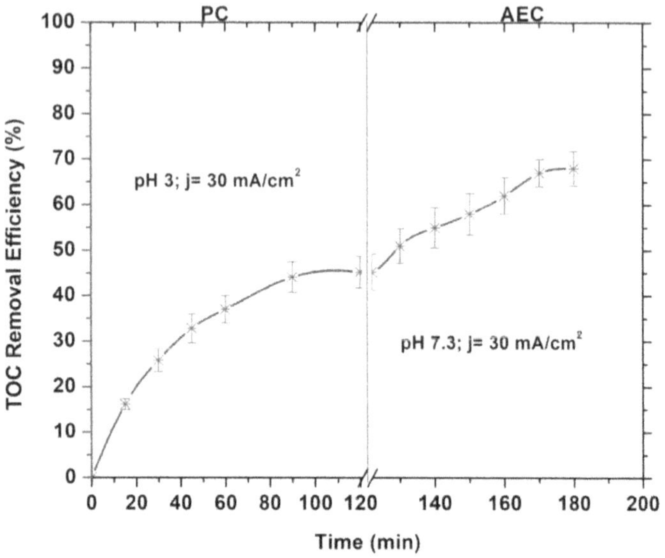

(a)

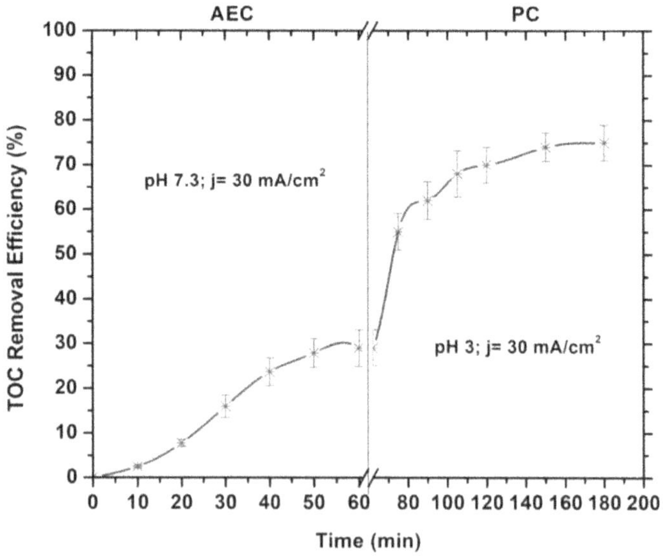

(b)

Fig. 4.7 Comparison of Sequential process for the leachate treatment (a) PC-AEC and (b) AEC-PC for the TOC removal

The front-loading coagulation conditions in the leachate treatment, stripes particulate and high-molar-mass organics and lowering the pool of radical scavengers, so the subsequently generated $HO^{\bullet}$ is consumed more productively on the residual dissolved fraction. Operationally, the AEC first order also mitigates bubble shielding and improves mass transfer for the oxidation stage, while aeration in AEC curtails Fe_3O_4 passivation on iron surfaces.

Overall, the sequence clearly matters; AEC-PC achieves higher TOC removal and exhibits a pronounced acceleration immediately after the stage switch, consistent with matrix conditioning by coagulation.

4.5 Sludge Quantification of Standalone and Sequential Process

Table 4.2. Sludge generated in the AEC, PC and Sequential AEC-PC process

Parameter	AEC Sludge	PC Sludge	Sequential AEC-PC	Sequential PC-AEC
Sludge volume (mL/L treated)	11.25	17.50	21.5	23.20
Wet mass (g)	12.25	18.66	22.8	24.86
TS per L of treatment (g/L)	0.4483	0.3713	0.604	0.7192
VS per L of treatment (g/L)	0.1463	0.1108	0.189	0.2274
Moisture (%)	96.34	98.01	97.34	97.11
TS (% of wet mass)	3.66	1.99	2.65	2.89
VS (% of TS)	32.63	29.83	31.29	31.6
TS conc. per L of sludge (g/L)	39.84	21.21	28.09	31.00
VS conc. per L of sludge (g/L)	13.00	6.33	8.79	9.80

FS conc. per L of sludge (g/L)	26.84	14.89	19.3	21.20
Apparent wet sludge density (g/mL)	1.089	1.066	1.0605	1.071

Sludge generation was quantified for standalone aerated electrocoagulation (AEC), peroxicoagulation (PC), and the two sequential orders (AEC-PC and PC-AEC) and is given in Table 4.2. For per litre (L) of leachate treated, the AEC-PC sequence generated a total solids (TS) yield of 0.604 g/L, compared with 0.4483 g/L for AEC and 0.3713 g/L for PC, and produced the largest wet-sludge volume among these three of 21.5 mL/L (wet-sludge mass 22.8 g/L). However, the withdrawn AEC-PC sludge was more dilute (TS = 2.65% of wet mass; moisture 97.34%; TS concentration 28.09 g/L; apparent density 1.0605 g/mL) than the AEC sludge. AEC yielded the most compact sludge (TS = 3.66%; moisture 96.34%; TS concentration 39.84 g/L; density 1.089 g/mL) at a smaller wet volume of 11.25 mL/L and wet-sludge mass of 12.25 g/L. PC generated the most hydrated sludge (TS = 1.99%; moisture 98.01%; TS concentration 21.21 g/L; density 1.066 g/mL) with 17.50 mL/L wet-sludge volume and 18.66 g/L wet-sludge mass. Volatile solids (VS) under these three conditions were 30–33% of TS (AEC 32.63%, PC 29.83%, AEC-PC 31.29%), indicating a predominantly inorganic, iron-rich sludge (fixed solids 67–70% of TS). Although PC exhibited a larger wet-sludge volume than AEC, its dry-solids yield per litre was lower (PC < AEC), indicating that PC produced more hydrated, lower-density gel-flocs (oxidation-gelled $Fe(OH)_x$ with oxygenated organics and entrained gas), whereas AEC produced denser $Fe(OH)_3$ sweep-flocs. Thus, PC still aligns with a degradation-dominant pathway (less TS), and the higher volume reflects hydration/settleability, not greater solids generation. Part of PC's higher wet-sludge volume arises from its longer runtime (120 min), which together with pH drift from 3 toward 4–5 produces fine, highly hydrated gel-flocs and greater gas entrapment, inflating volume despite a lower dry-solids yield than AEC (60 min).

PC-AEC (sequential) showed the highest TS capture overall, 0.719 g/L, with the largest wet-sludge volume (23.20 mL/L) and wet-sludge mass (24.86 g/L). The withdrawn PC-AEC sludge remained moderately dilute (TS = 2.89%; moisture 97.11%; TS concentration 31.00 g/L; density 1.071 g/mL; VS 31.60% of TS). The order effect explains this outcome: the first-stage PC partially oxidizes/opens organics and generates fine Fe(III) nuclei; the second-stage AEC at near-neutral pH then sweeps these colloids and dissolved species as $Fe(OH)_3$ flocs, raising dry-solids captured per litre while maintaining a hydrated structure (volume ↑). As with AEC-PC, the sequential HRT (180 min) and gel-floc/gas-entrainment effects increase wet volume more than TS concentration. The iron-oxide-rich sludge from AEC (and retained into the sequences) can also be left in the effluent to catalyze the subsequent PC step, lowering operating cost. Further sludge generated from electrochemical process can be utilized in the following disposal methods like manufacturing of bricks which increases the compressive strength of brick [155,240,241] in the production of colored glass (gave color to the glass) [155,242] etc.

4.6 Energy Consumption

Energy Consumption can be explained as the amount of energy that can be directly used per amount of product. It is used as an indicator to evaluate or measure the performance of energy efficiency [243]. The energy consumption of the sequential AEC-PC process was calculated by using the following Eq. (4.44)

Energy Consumption (KWh/Kg of COD removed) =

$$\frac{VIt \times 10^3}{60(COD)_o - (COD)_{t)} \times treated\ volume(L)} \tag{4.44}$$

where V is the cell voltage (V), I is the current (A), t is the electrolysis time (min), COD_o and COD_t are the initial COD concentration demands and COD concentration at time t (mg/L), respectively [145].

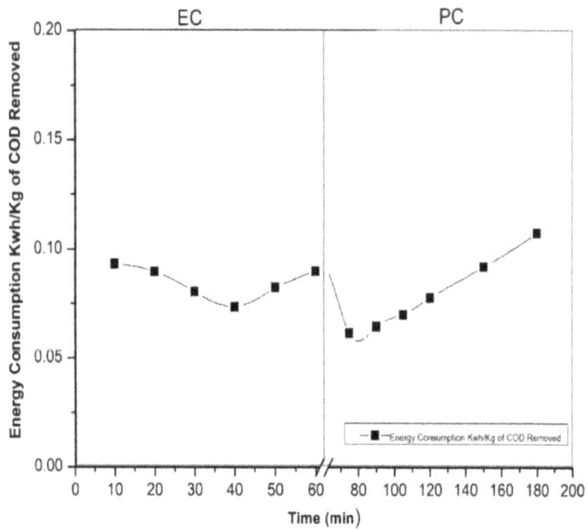

Fig. 4.8. Energy Consumption (kWh/Kg of COD removed) (Operating conditions applied voltage of 3 V, Current density 30 mA/cm^2, electrolysis time of 180 min)

From Fig. 4.8 maximum energy consumption of 0.107 kWh/Kg of COD removed was reported after 180 minutes of reaction time and minimum consumption of 0.061 KWh/Kg of COD removed after 75 minutes of electrolysis for AEC-PC and for PC-AEC the maximum energy consumption of 0.12 kWh/Kg of COD removed was reported after 180 minutes of reaction time and minimum consumption of 0.043 KWh/Kg of COD removed after 15 minutes of electrolysis [145]. Upto 40 mins there is an increase in COD removal in the case of AEC, further with increase in electrolysis time the COD removal decline, due to the increase in temperature of bulk liquid [3], whereas the increase in time increases the production of ferric ion complexes [116] which enhances the energy consumption/Kg COD removed. From Fig. 4.4 it's evident that there is an increase in COD reduction in the initial 20 min of peroxicoagulation treatment due to direct oxidation resulted in COD degradation followed be first order kinetics and after that there is a decline in COD reduction resulted in the increased energy consumption/Kg of COD removed.

4.7 Synergistic Index

Total efficiency of the process may be expressed as a sum of the removal efficiencies of aerated electrocoagulation and peroxicoagulation. Thus, the synergism of both processes was assessed by using a synergistic index (SI) expressed below [244].

$$\text{Synergistic index, SI}=\frac{RR}{(RR_{electrocoagulation}+RR_{peroxicoagulation})}$$

RR- Removal rate of COD for the sequential process

$RR_{aerated\ electrocoagulation}$- Removal rate of COD for standalone aerated electrocoagulation treatment

$RR_{peroxicoagulation}$ – Removal rate of COD for standalone peroxicoagulation treatment

The removal rate of COD for the sequential treatment was 78%. The standalone removal rate of AEC and PC was found to be 26.2% and 37.3% respectively. The synergistic index for the sequential process has been computed as 1.22 based on the removal efficiency of observed for the standalone process and for the sequential process. An additive effect occurs when the combined process effect is equal to the sum of the sole process effect when the synergistic index (SI) is 1, a synergistic effect occurs when the sequential process effect is significantly greater than the sum of the sole process effect, and antagonism occurs when SI is less than 1 [245]. The synergistic index of 1.22 indicates a high synergistic effect on the sequential AEC and PC process for the treatment of landfill leachate.

4.8 Summary

In this study, two standalone processes aerated electrocoagulation and peroxicoagulation process was used for the treatment of landfill leachate. Sequential AEC and PC process was also adopted for the treatment of leachate. The elimination of organics and nitrogen pollutant loads was substantially enhanced by the suggested sequential AEC and PC process. These are the key conclusions drawn from the study,

The AEC method exhibits a removal of 26.2% COD, 28.9 % TOC, 25.6% NH_4–N, and 23.6% TN with a current density of 30 mA/cm^2, pH 7.3 and reaction time of 60 min. Whereas PC shows a removal of 37.3% COD, 45.7 % TOC, 37.3% NH_4–N, and 43.7%

TN with current density of 30 mA/cm^2, pH 3 and reaction time of 120 min. Studies done by Abdullah et al. [246] for the treatment of landfill leachate using electrocoagulation removed around 39% of COD and 12% of TOC at 90 mins of electrolysis time due to the limited effect on soluble organics. In the current study, the sequential AEC and PC process has shown an efficient removal of 78% COD, 75% TOC, 63% NH_4-N, and 68% TN with a current density of 30 mA/cm^2, pH value 7.3 for AEC and for PC the current density was 30 mA/cm^2, pH 3 and with a total reaction time of 180 min.

The AEC and PC process performs effectively for the treatment of landfill leachate in a sequential manner. It shows the potential of sequential electrochemical methods having the ability to treat complicated wastewater. For the treatment of landfill leachate, the synergetic index of the sequential AEC-PC process has been calculated. The synergetic index was determined to be 1.22, which is higher than one and which indicates that the sequential process had a significant synergetic effect. It is visible from the reported results that the efficiency of the sequential AEC and PC process is comparably high and can be used as an effective process for the treatment of landfill leachate.

Chapter 5

Wood Waste-Derived Cathode for EAOP: Synthesis & Characterization

5.1 General

Landfills are widely used for waste disposal due to their economic advantages, but poorly managed sites generate leachate that contaminates soil and water. As landfills stabilize, leachate becomes resistant to biodegradation, characterized by a low BOD/COD ratio ($<$ 0.1), high ammoniacal nitrogen, and persistent organic pollutants, limiting conventional biological treatments. Advanced oxidation processes, particularly electro-Fenton and peroxicoagulation, provide promising alternatives by producing strong oxidants like hydroxyl radicals in situ. Cathode materials are critical for hydrogen peroxide generation and overall efficiency. While carbon-based electrodes such as graphite are commonly used, they rely on non-renewable sources. Biomass-derived cathodes, especially from lignocellulosic biomass like wood waste, represent a renewable, cost-effective alternative with high surface area and porosity. This chapter focuses on the development and characterization of a BDC, prepared as per the methodology in Chapter 3, and its application as a cathode in the peroxicoagulation treatment of stabilized landfill leachate.

5.2 Characterization of Biomass Derived Cathode

To have a better understanding of the biomass derived cathode TGA-DSC, SEM, XRD, Raman spectroscopy, Fourier transform infrared spectroscopy and EDX was carried out, and the results are discussed as below.

5.2.1 TGA-DSC of Biomass Derived Cathode

The optimum pyrolysis temperature profile was selected using TGA-DSC analysis of waste wood in an inert environment. The findings are shown in Fig. 5.1. The TGA analysis of the raw wood has shown three stages of weight loss at 25 °C-225 °C, 225 °C-375 °C, and 375 °C-475 °C as shown in Fig. 5.1(a).

At 120 °C, an endothermic peak was seen in the first region, which was related to moisture evaporation and a 10% weight loss was observed. Then, between 225 °C and 375 °C, a greater weight loss (70%) occurred, which was mostly attributed to hemicellulose degradation because of its fragile amorphous structure [247]. Lignin, cellulose and hemicellulose make up the majority of lignocellulosic biomass. The most unstable of the three components is hemicellulose which primarily decomposes between 220 and 315 °C [248]. According to earlier studies Shang et al. and Chowdhury et al. [249,250] the weight loss occurs between 375 and 475 °C was around 10%, which was attributable to the lignin and cellulose's initial carbonization process, which eventually produces a graphitic structure inside the biochar matrix. The DSC line in Fig. 5.1(b), shows the thermal condensation of the waste wood. Assuming a suitable conducting graphitic structure was generated from 500 °C, the pyrolysis temperature range was chosen from 500 °C to 900 °C. According to Yang et al. [248] cellulose has superior heat stability and mostly decomposes between 315 and 400 °C. Because of its more convoluted structure and complex degradation mechanism, lignin can be thermally degraded at a wide range of temperatures (150-900 °C). Waste wood produces low carbon biochar with limited conductivity at temperatures below 500 °C, making it unsuitable for use as an electrode material.

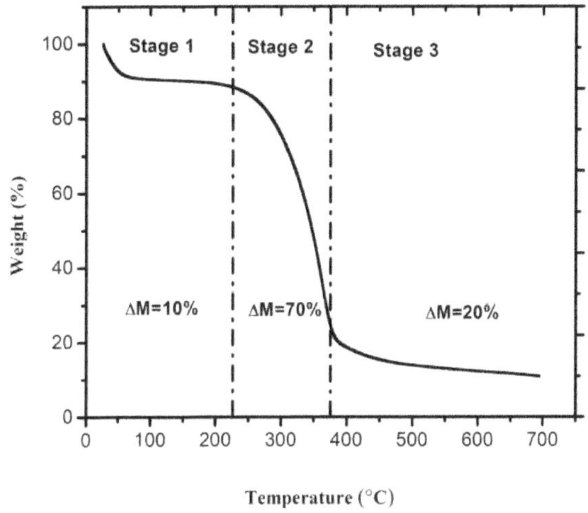

(a)

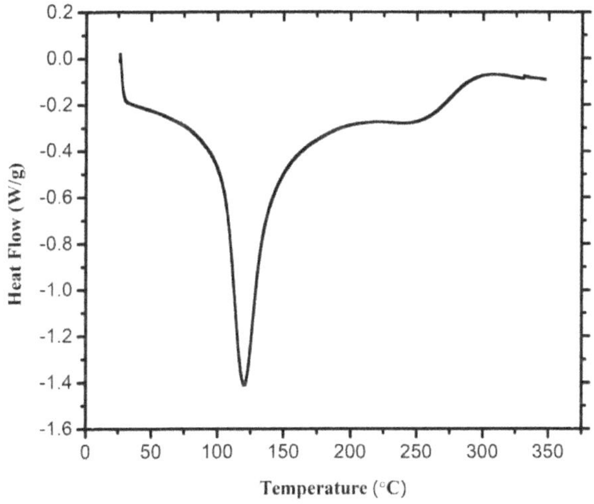

(b)

Fig. 5.1 (a) Thermogravimetric analysis and (b) Differential scanning calorimetric of BDC

5.2.2 BET Analysis of Biomass Derived Cathode

The total surface area of the raw wood and pyrolyzed wood (BDC) was determined using BET analysis, by determining the quantity of adsorbed N_2.

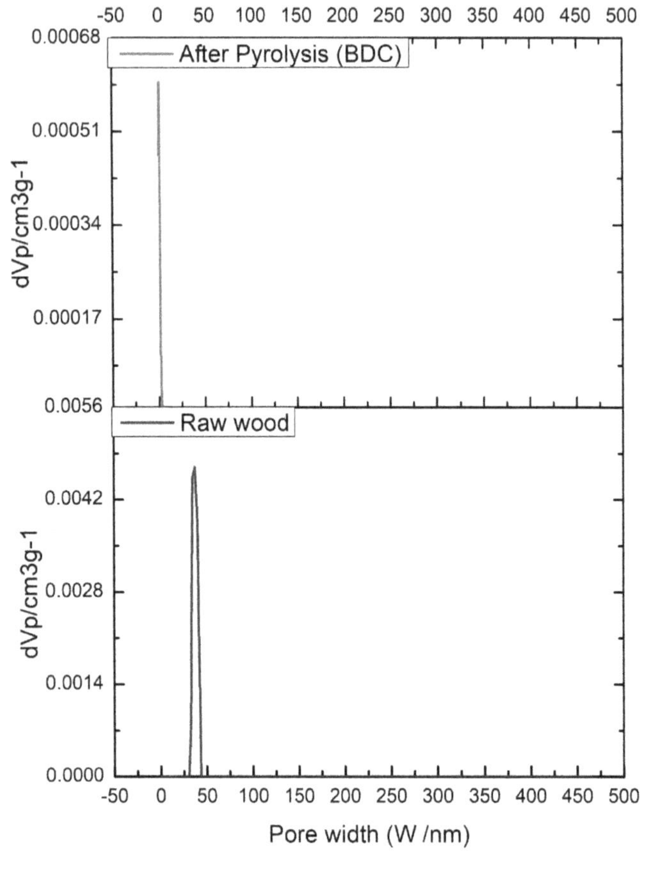

Fig. 5.2 Pore size distribution of raw and pyrolyzed wood (BDC)

As a result of the organic matter in the feedstock decomposing and volatilizing with the increase in the pyrolysis temperature and time the biochar's BET surface area increased. The structure of biochar is composed of micropores (< 2 nm), mesopores (2–50 nm), and macropores (> 50 nm). Mesopores are in charge of rapid diffusion, macropores might improve the transport of chemical substances and micropores could offer adsorption sites and contribute to the high specific surface area [251]. According to BET surface analyses an increase in surface area occurs with the increase in temperature. The surface area observed was 1.7687 m^2/g and 7.8807 m^2/g for raw wood and for the pyrolyzed wood at 900 °C respectively. This may be due to the fact that the unstable components continue to break down into volatile molecules including CO_2, H_2, methanol, and tar as the pyrolysis temperature rises. The overflow of these volatile molecules blows away the material obstructing the pores, forming new active sites and more porous structures [252, 253]. The pore distribution results are shown in Fig. 5.2. Majority of the pores are in the mesopores range for the raw wood and after the treatment there is an increase in the micropores range. After the treatment 98.2% of the pore volume was contributed by the micropores structure whereas initially 100% pore volume was contributed by the mesopores. Active sites and biochar reactivity are strongly correlated, active sites require a large accessible surface area. Applications of biochar are also greatly influenced by surface area and porosity.

5.2.3 XRD Analysis

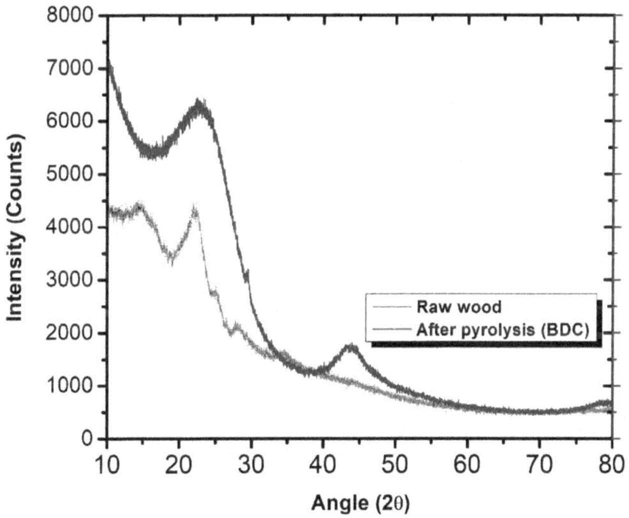

Fig. 5.3 XRD analysis of biochar raw wood and BDC

The diffractograms (Fig. 5.3) evidence the transformation from a cellulose-dominated structure in raw wood to turbostratic graphitic carbon after pyrolysis. In the BDC, the broad maximum at $2\theta \approx 27.9°$ is assigned to the (002) reflection of disordered/stacking-faulted graphitic layers [254], while the band at 42.7° corresponds to the in-plane (101) reflection of hexagonal graphite [179]. These two features (002) and (101) are the canonical signatures of turbostratic, graphene-layer stacks in biochars and biocarbons, and they are consistently reported for waste-wood/bamboo-derived graphitic biochars used as electro-Fenton cathodes [179,254,255]. In contrast, the raw wood does not show the high angle graphite reflections; instead, its pattern is dominated by the broad cellulose-I background that is known to diminish upon carbonization or graphitization. The disappearance of cellulose features together with the emergence and sharpening of the (002) / (101) bands confirm progressive graphitization in the BDC [256].

5.2.4 SEM Analysis

The SEM images of raw and pyrolyzed wood are shown in Fig. 5.4. The pyrolysis of wood at 900 °C of temperature developed a permeable structure, which are very much adjusted like a miniature channel such structures are not found in the raw wood (Fig. 5.4(a)).

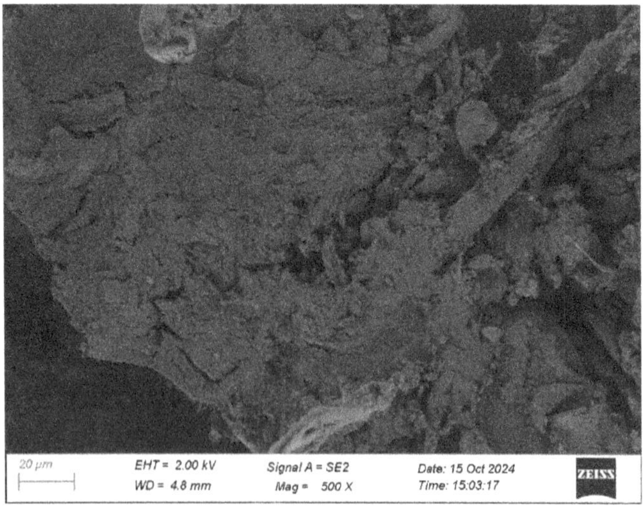

(a)

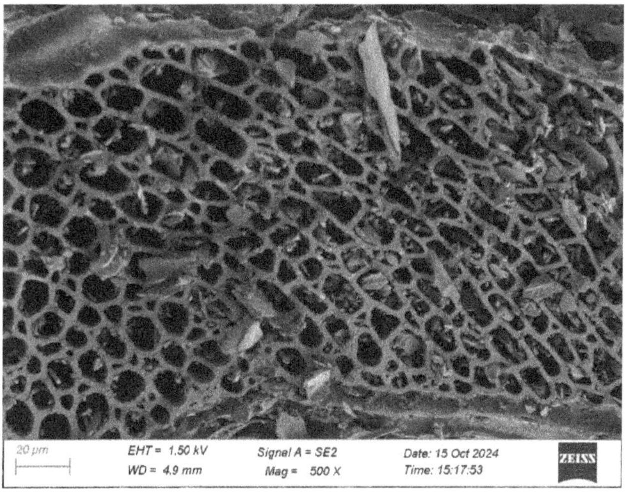

(b)

(c)

Fig. 5.4 SEM image of (a) raw wood (b) pyrolyzed wood (BDC) (c) graphite electrode

As seen in Fig. 5.4(b), the structure looks more like a cross sectional and miniature channel walls. The aligned microfibers penetrated into the substrate completely, creating channels that joined to construct the intended mass transport pathways [257]. This open microchannels developed on the BDC facilitates the diffusion of O_2 [258], which in turn provides adequate number of dynamic locales for the ORR. This facilitation promotes the production of H_2O_2 required for the EAOP reaction [254]. Pyrolyzed wood has shown more porous like structure compared to that of graphite as shown in Fig. 5.4(c).

5.2.5 FTIR Analysis

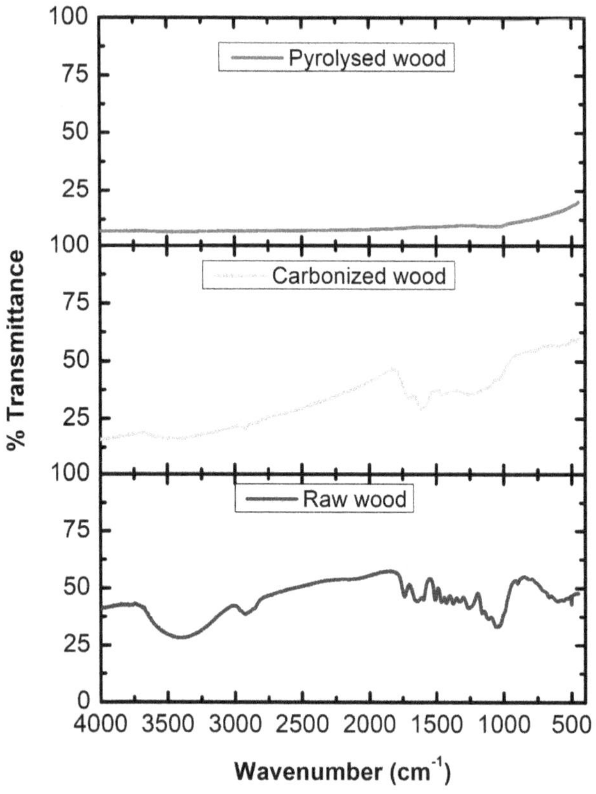

Fig. 5.5 FTIR spectra of Raw wood, Carbonized wood and Pyrolyzed wood (BDC)

The characteristic FTIR analysis of wood is performed at different temperature condition (raw wood, 300 °C carbonized wood and 900 °C pyrolyzed wood) and are shown in Fig. 5.5. It can be observed from the Fig. 5.5 that FTIR signals progressively lose their characteristics peaks with increasing the temperature from raw wood stage (Fig. 5.5).

Dehydration is the most significant reaction during pyrolysis [259], as evidenced by one of the most substantial reduction in the wavenumber range of 3600-3200 cm^{-1}, which is related to the hydrogen bond O–H stretching vibration [256]. Decreases in aliphatic C-H groups (3000–2800 cm^{-1}), alkyl sulfonate functionalities (1150 cm^{-1}), and C-O functionality (about 1030 cm^{-1}) are all associated with aromatic condensation. Additionally, as the temperature rises, the broad band with sub-bands between 1600 and 1000 cm^{-1} (such as the in-plane C-H bending vibrations interact with the different aromatic ring C-C vibrations) significantly decreases, indicating the progressive aromatization of the formed carbonaceous material [260]. All characteristic signals are significantly reduced during pyrolysis at 900 °C, signifying the loss of most functional groups and the beginning of graphitization because graphite lacks distinctive infrared bands in the wavenumber under investigation [255,260].

5.2.6 EDX Analysis

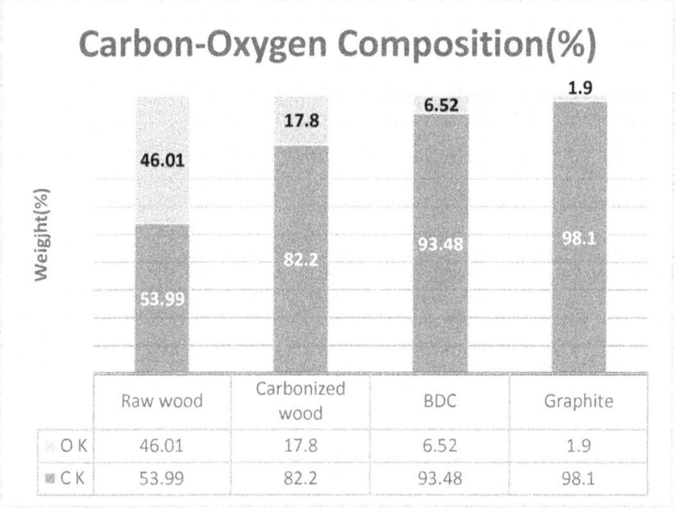

Fig. 5.6 EDX analysis of raw wood, carbonized wood, BDC and graphite electrode

When EDX analysis was performed on the raw wood, carbonized wood, BDC and graphite electrode used for the study, the results reveal a decrease in the percentage of oxygen and an increase in the carbon content in relation to the treatment temperature. The carbon content of the raw wood was 53.98% and it increased to 82.2% after carbonization at 300 °C and 93.48% after pyrolyzed at 900 °C. The oxygen content dropped to 5.71% after pyrolysis when compared to that of 46.01% of raw wood, as shown in Fig. 5.6.

The decline in oxygen and increase in carbon is due to the dominant reactions like carbonization and deoxygenation occurred at 900 °C [261]. The carbon content of BDC electrode was comparable with the carbon content of graphite electrode which is 98.1% as in Fig. 5.6. The BDC contains 93.48% carbon, slightly lower than the 98.1% carbon content found in commercial graphite. This modest difference in carbon content may lead to slightly reduced conductivity [262]; however, the presence of residual heteroatoms (e.g., oxygen) could enhance surface functionality, thereby potentially benefiting the electrochemical activity despite the lower carbon purity.

5.2.7 Raman Analysis

The Fig. 5.7, shows the Raman spectra of raw wood and pyrolyzed wood. The different bonds and lattice vibrations of the wood materials are reflected in the Raman spectrum, there are two main Raman peaks D and G in the Raman spectra of carbon materials [263, 264]. In general, the G peak is positioned at the range of 1550-1615 cm^{-1} and it corresponds to the stretching of carbon sp^2 bonds [265,266], whereas the edge disorder is related to D peak which ranges between 1300-1400 cm^{-1} (breathing mode, A1g-mode) [267]. The presence of G & D peaks in the respective range for the pyrolyzed wood indicated that BDC has layered graphite structure [266]. Bands and signals in the Raman spectrum correspond to the stretching of various bonds and vibrations in the lattice. The intensity of the D and G peaks (I_D/I_G) is divided to get the ratio of disordered to graphitized structures in the carbon material being tested. The degree of disorder and graphitization in the carbon material is reflected by this ratio. The increased disorder degree would cause more electrons to flow across the π-electron system, improving BDC's conductivity [268,269].

The Raman analysis of the pyrolyzed wood evidenced that there was a drastic increase in the defected graphite like carbon. The increase in defected graphite like carbon increases the conductivity of the BDC [259]. The increase in the conductivity of BDC material find its potential application in the field of electro chemical process. As discussed, the increase in the treatment temperature led to a change in the carbon structure which further leads to the development of graphitic structure and also EDX analysis showed reduction of organic component of wood with the increase of carbon concentration related to the conductivity. Moreover, microstructure evolution also led to development of conductivity of biochar [259].

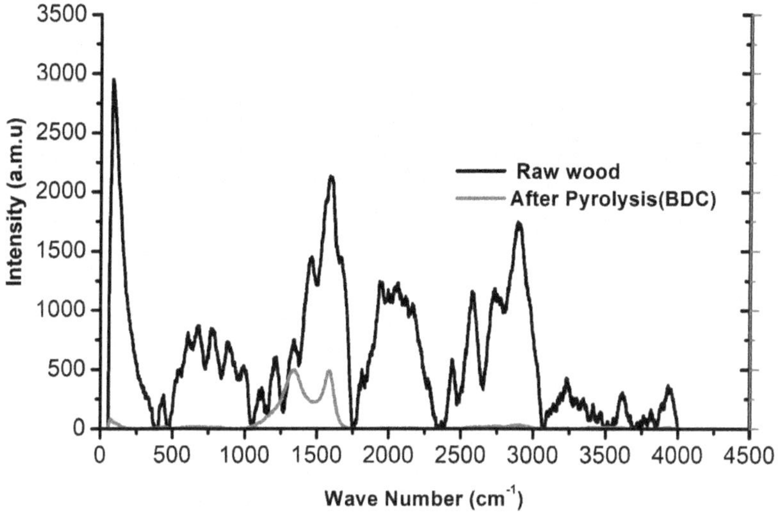

Fig. 5.7 Raman characterization for waste wood biochar in function of pyrolysis temperature

5.2.8 Wettability of Electrode

The wettability of the electrode is measured using contact angle of the electrode. If the contact angle is less than 90° it is hydrophilic and if the angle is greater than 90° it is hydrophobic. The contact angle of graphite (Fig. 5.8(a)) and BDC electrode (Fig. 5.8(b)) is 122° and 17° respectively. It was observed that increase in the hydrophilicity of the BDC cathode, which was confirmed by the lowest contact angle of the material. As a consequence, the amount of dissolved oxygen in the aqueous solution in contact with the cathode surface, favours the H_2O_2 generation [181,254,270].

(a)

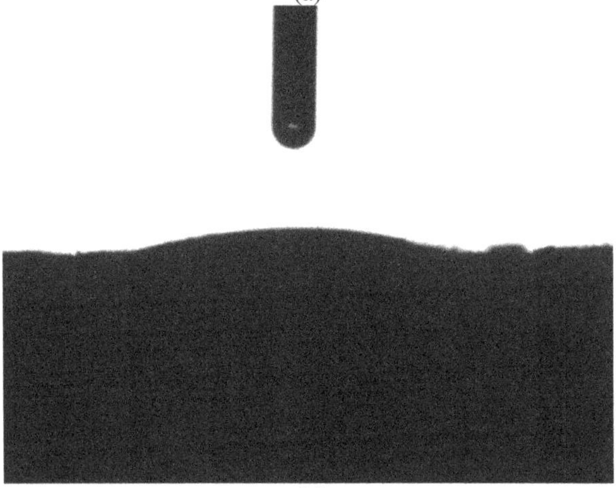

(b)

Fig. 5.8 Contact angle image of (a) graphite and (b) BDC

5.2.9 Electrochemical Impedance Spectroscopy and Cyclic Voltammetry Analysis

The electro-active surface area was estimated by using Cyclic voltammetry measurements (Fig. 5.9(a)) in a mixed solution of 5 mM $K_3Fe(CN)_6$ and 1 M KCl at a scan rate of 1 mV/s and with a potential range of 1.2 V to −1.4 V vs Ag/AgCl. According to the Randles-Sevcik equation (Eq. 5.1) the electroactive surface areas of graphite electrode and BDC electrode were 0.89 cm^2 and 1.138 cm^2 respectively.

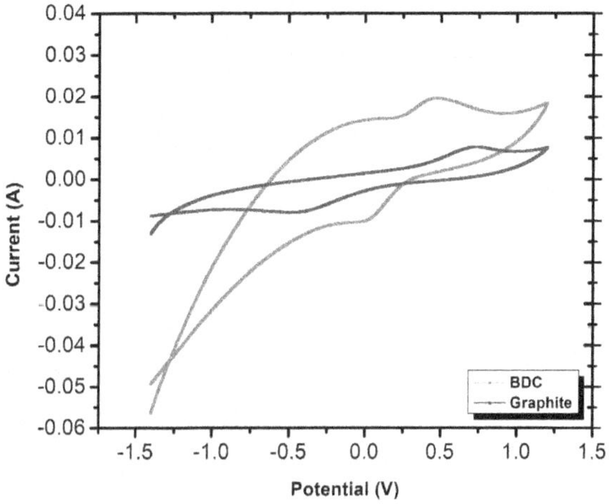

(a)

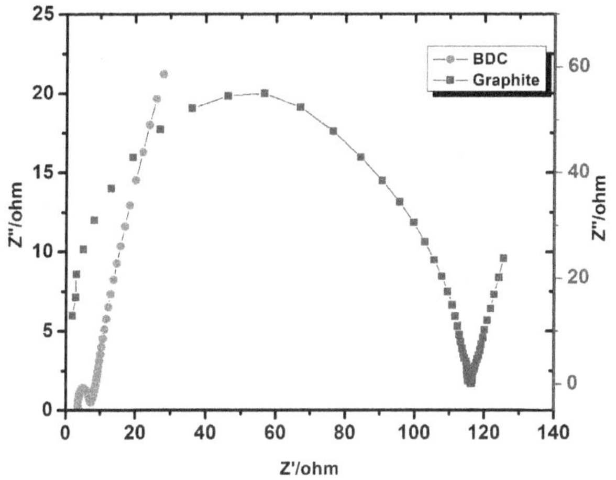

(b)

Fig. 5.9 (a) CV scans of BDCs electrode and graphite electrode, (b) Nyquist plots of BDCs electrode and graphite electrode

The electro active surface area of BDC electrode was 1.27 times higher than the commercial graphite electrode.

$$I_p = 2.69 \times 10^5 \times AD^{1/2} \times n^{3/2} \times \gamma^{1/2} \times C \quad (5.1)$$

Where I_p - current maximum (Amps); n - number of electrons transferred in the redox reaction; A - electrode area in (cm^2); D - diffusion coefficient in (cm^2/s); C - concentration in ($Mole/cm^3$); γ- scan rate in (V/s).

Cell & geometry: Custom glass-vial cell, with flat 1.0 × 1.0 cm graphite or BDC coupons mounted vertically; exposed area = 1.0 ± 0.1 cm^2. Counter: Pt mesh; reference: Ag/AgCl (1 M KCl) placed 2–3 mm from the working face via a Luggin tip. Temperature 25 ± 1 °C; Cleaning/conditioning: DI rinse → isopropanol wipe → 10 background cycles before recording data.Electrolyte & gases: 5 mM $K_3[Fe(CN)_6]$ + 1.0 M KCl; N_2 background followed by O_2.

Increase in electroactive sites improved the oxygen reduction reaction [271]. The electro active surface area of the electrodes used in the electrochemical process place a vital role in the oxygen reduction reaction. The production of H_2O_2 in the aqueous system depends on the ORR thus the increase in the electroactive surface area enhances the electro chemical production of H_2O_2.

EIS studies were also carried out in order to clarify the electrodes' ability to transmit electrons. The Nyquist plots of graphite electrode and BDC electrode are displayed in Fig. 5.9(b). As shown in the Nyquist plot (Fig. 5.9(b)), the BDC electrode exhibits a smaller semicircle diameter compared to graphite, indicating lower charge transfer resistance. The extracted charge transfer resistance values are 50 Ω for BDC and 120 Ω for graphite, confirming enhanced charge transfer kinetics of the BDC. The lesser charge transfer resistance indicates the BDC electrode has a smaller impedance for charge transfer [272]. EIS results shows that the charge transfer ability of BDC is stronger than that of graphite electrode.

5.3 Peroxicoagulation Process for the Treatment of Leachate Using Biomass Derived Cathode

Degradation of organic pollutants are significantly influenced by the production and activation of H_2O_2. A $2e^-$ ORR reaction at the cathode produces H_2O_2 electrogeneration by the triple phase (gas, liquid, and solid) interfacial interactions. The generic two-electron oxygen reduction is represented by Eq. (5.2), but the pathways vary depending on the medium. In a basic aqueous environment the following reaction takes place Eqs. (5.2-5.5) [253].

$$O_2 + H_2O + 2e^- \rightarrow H_2O_2 \quad (5.2)$$

$$O_2 + H_2O + 2e^- \rightarrow HO_2^- + OH^- \quad (5.3)$$

$$O_2 + H_2O + e^- + * \rightarrow * OOH + OH^- \quad (5.4)$$

$$* OOH + e^- \rightarrow HO_2^- + * \quad (5.5)$$

where * represents the active sites. The generated HO_2 can be further reduced by another two-electron reduction process or decomposed into OH^-, as shown in Eqs. (5.6) and (5.7).

$$HO_2^- + H_2O + 2e^- \rightarrow 3OH^- \quad (5.6)$$

$$2HO_2^- \rightarrow 2OH^- + O_2 \quad (5.7)$$

In acidic environment, the reaction will be as per Eqs. (5.8)-(5.10)

$$O_2 + 2H^+ + 2e^- \rightarrow H_2O_2 \quad (5.8)$$

$$O_2 + H^+ + e^- + * \rightarrow * OOH \quad (5.9)$$

$$* OOH + H^+ + e^- \rightarrow * + H_2O_2 \quad (5.10)$$

The generated H_2O_2 can also be further reduced by another two-electron reduction process or chemical disproportionation, as shown in Eqs. (5.11) and (5.12).

$$H_2O_2 + 2H^+ + 2e^- \rightarrow 2H_2O \quad (5.11)$$

$$2H_2O_2 \rightarrow 2H_2O + O_2 \quad (5.12)$$

Particularly at high concentrations of H_2O_2, the produced H_2O_2 is vulnerable to the catalytic degradation in the alkaline medium, whereas in the acidic medium the H^+ catalytic reaction promotes the generation of H_2O_2 Eq. (5.12). Therefore, the majority of the current research has been on the electrolytic production of H_2O_2 in media with an acidic pH usually about 3 [253,273].

5.3.1 Effect of pH on the Treatment of Leachate Using Biomass Derived Cathode

Electrochemical treatment was carried out using peroxicoagulation process using biochar as cathode. The operating conditions were decided based on the peroxicoagulation process based on graphite cathode and as discussed in Chapter 4. As shown in Eqs. (5.13) and (5.14), the electrolysis of the cell produces hydrogen peroxide and ferrous ions during the peroxicoagulation process. The organic contaminants are eliminated through coagulation and degradation, by combining the electro-Fenton and electrocoagulation processes, as shown in Eqs. (5.15) and (5.16) [113].

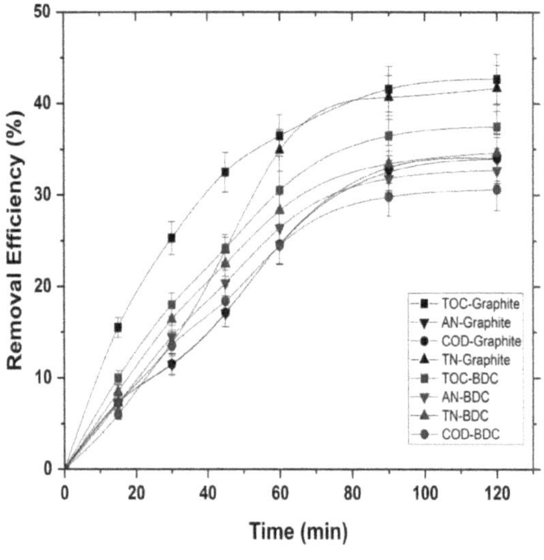

(a)

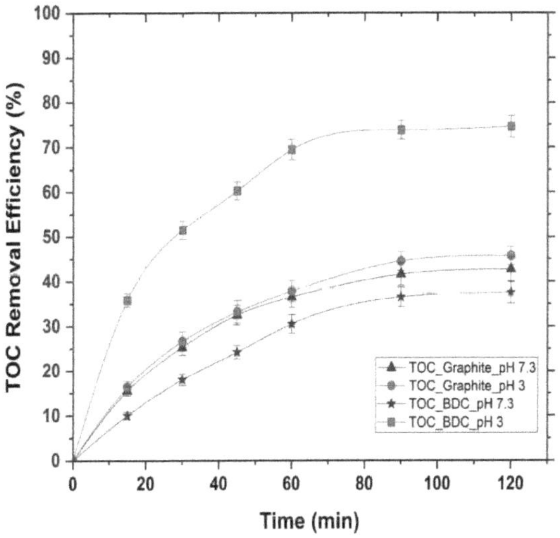

(b)

Fig. 5.10 Removal of (a) AN, TN, COD, TOC using BDC and graphite as cathode at pH 7.3 and (b) TOC using BDC and graphite as cathode at pH 3 and 7.3

$$2Fe \rightarrow 2Fe^{2+} + 4e^- \quad (5.13)$$

$$O_2 + 2H^+ + 2e^- \rightarrow H_2O_2 \quad (5.14)$$

$$Fe^{2+} + H_2O_2 \rightarrow Fe^{3+} + HO^- + HO^\bullet \quad (5.15)$$

$$2Fe^{2+} + 4HO^- \rightarrow 2Fe(OH)_2 \quad (5.16)$$

The pH has a significant impact on the oxidation reaction because when the pH rises over 3, the H_2O_2 becomes unstable, forming molecular oxygen. At solution pH below 3, Fe^{2+} (i.e., $Fe(H_2O)_6^{2+}$) and Fe^{3+} (i.e., $Fe(H_2O)_6^{3+}$) are the most common forms of iron among the several hexa-aqua complex species. The iron species at the core is covalently bound to six water molecules. One of the oxygen lone electron pairs is used by the water molecule to establish a link. As in Eqs. (5.17) and (5.18), this ferric complex is further hydrolysed. In a similar vein, complex formation also occurs in ferrous ions. These complexes change into insoluble complexes such as $[Fe(H_2O)_8(OH)_2]^{4+}$, $[Fe_2(H_2O)_7(OH)_3]^{3+}$, $[Fe_2(H_2O)_7(OH)_4]^{5+}$, etc., as the pH of the solution rises above 3.

Under these circumstances, coagulation-based pollutant removal dominates the degradation process. It has been demonstrated experimentally that the ideal pH range for the electrocoagulation of organic contaminants employing iron anode is between 6 and 8. Additionally, ferrous ions oxidize to ferric ions at pH values higher than 4 when oxygen is present as shown in Eq. (5.19). At pH values greater than 4, this slows down the rate of Fenton's reaction. Because ferrous and ferric ions are more prevalent at pH 3, where the Fenton's reaction proceeds at faster rates. The anodic oxidation of water in the EF process keeps this pH close to the starting point [274].

$$[Fe(H_2O)_6]^{3+} + H_2O \rightarrow [Fe(H_2O)_5OH]^{2+} + H_3O^+ \quad (5.17)$$

$$[Fe(H_2O)_5OH]^{2+} + H_2O \rightarrow [Fe(H_2O)_4(OH)_2]^{2+} + H_3O^+ \quad (5.18)$$

$$Fe^{2+} + O_2 \rightarrow Fe^{3+} + O_2^{\bullet -} \quad (5.19)$$

The treatment process is carried out at pH 3 and the maximum removal efficiency of TOC observed was 77% as shown in Fig. 5.10(b). During the first half an hour of the PC process, the rate of removal is higher; however, as electrolysis increases, the rate of removal efficiency declines. As the pH of the leachate was 3 and the electrolytic generation of iron at low pH is Fe^{2+}, initially the dominant mechanism maybe the electro-Fenton process as shown in Eq. (5.15). As the reaction time increases there is a change in the pH of the leachate and thereby, there may be a shift in dominant mechanism further leading to a coagulation cum degradation mechanism as shown in Eq. (5.16) [116]. As the electrolysis time increased beyond 90 min, the removal of pollutants from the leachate gets almost saturated. The extended electrolysis period beyond 90 minutes resulted in a greater abundance of soluble iron ions. Consequently, electrode poisoning occurred, leading to a reduction in the catalytic activity of the anode.

In the previous Chapter (4), as discussed, the treatment of leachate with graphite electrode as cathode, the maximum removal efficiency of COD, TOC, AN, and TN, were 34%, 42.7%, 34%, and 41.7%, respectively, at a pH of 7.3, a current density of 30 mA/cm^2, and an operating time of 120 minutes as shown in Fig. 5.10(a). Similarly, 45.7% of TOC removal was observed at pH 3 (Fig. 5.10(b)). The TOC removal from leachate with graphite and BDC as cathode electrodes were 42.7% and 37.5% respectively. The graphite electrode poses a little higher removal efficiency compared to that of the BDC electrode in the pH 7.3 where the dominant mechanism is electrocoagulation rather than oxidation. But when the pH is decreased to 3, the TOC removal efficiency with graphite and BDC as cathode electrodes were 45.7% and 77% respectively. When pH was adjusted to 3 from 7.3 the TOC content decreases due to the fact that precipitation of humic acid compounds in acidic environment [222]. Further the humic acid and fulvic acid present in the leachate can be removed using the electrochemical oxidation. Moreover, at pH around 3, the iron species predominantly existed as ferrous and ferric ions. This accelerates the pace of Fenton reactions and consequently the breakdown of contaminants. As the pH increases from 3, the concentrations of ferrous and ferric ions decrease [31]. The electrochemical characteristics like cyclic voltammetry and electrochemical impedance spectroscopy shows that the BDC electrode has advantage over graphite electrode in terms of electroactive surface area and lower impedance. The electroactive surface areas of graphite electrode and BDC electrode were 0.89 cm^2 and 1.138 cm^2 respectively, i.e., the electro active surface area of BDC electrode is 1.27 times higher than the commercial graphite electrode. The lower impedance and higher electroactive surface area compare to graphite electrode result in a strong charge transfer ability for BDC electrode which favours the ORR resulted in the production of H_2O_2 [271] which favours the Fenton reaction and thereby increases the hydroxyl radical generation which enhances the treatment efficiency. The microporous structure of BDC facilitates higher surface area and active sites, enhancing H_2O_2 generation and its diffusion. This structural advantage works synergistically with Fe^{2+} catalysis, leading to a significant enhancement in oxidation efficiency during leachate treatment.

5.4 Sequential Treatment of Leachate Using Biomass Derived Cathode

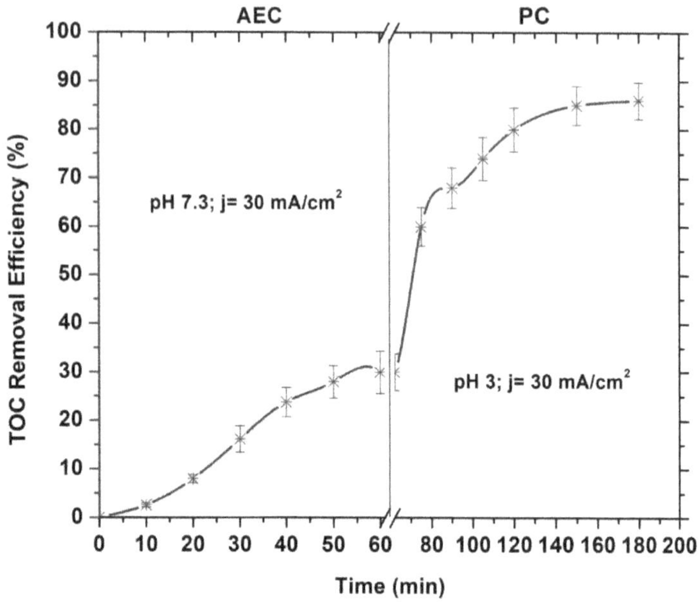

Fig. 5.11 Sequential AEC–PC process for the removal of TOC (AEC: pH 7.3, current density 30 mA/cm^2, electrolysis time of 60 min. PC: pH 3, Current density 30 mA/cm^2, electrolysis time of 120 minutes).

A two-stage electrochemical treatment was carried out to evaluate the complementary strengths of electrocoagulation and radical oxidation for stabilized landfill leachate. Stage I involved AEC, operated at near-neutral pH, to destabilize colloids, sweep-floc hydrophobic organics, and pre-remove particulate and adsorptive fractions while limiting iron passivation through aeration. Stage II employed the peroxicoagulation process at acidic pH to mineralize the remaining dissolved organics via in-situ H_2O_2 generation at a BDC electrode and catalytic Fe^{2+}/Fe^{3+} cycles. The BDC was selected to maximize the two-electron oxygen reduction reaction through its high micro/mesoporosity, abundant defect sites, and favorable wettability, thereby enhancing local H_2O_2 availability without chemical dosing.

Stage I (AEC) was conducted for 60 min at pH 7.3 with air sparging to sustain hydrodynamics and mitigate anode passivation. The suspension was then acidified to pH 3.0 and Stage II (PC) was carried out for an additional 120 min at the same current density, with the BDC serving as the cathode. No external oxidant or coagulant was added; iron was supplied from the sacrificial anode. TOC removal in the sequential AEC–PC process is shown in Fig. 5.11. During AEC (0–60 min), TOC removal increased steadily to 30%, reflecting the preferential elimination of particulate, colloidal, and adsorptive fractions. Upon switching to PC (60–180 min), a pronounced inflection occurred TOC removal jumped to 60% within the first 15 min and continued to rise, reaching 86% by 90 min of the PC stage.

5.5 Comparison of Sequential Treatment of Leachate Using Graphite and Biomass Derived Cathode

Sequential AEC–PC experiments were conducted at an identical current density (30 mA/cm^2). AEC was operated for 0–60 min at pH 7.3 using iron as both anode and cathode (Fe–Fe) followed by PC for 60–180 min at pH 3 using an iron anode while varying the cathode between graphite and the BDC. The results are shown in Fig. 5.12. As expected from the identical Fe–Fe configuration, the AEC stage exhibited indistinguishable performance in both cases achieving 30% TOC removal by 60 min. The profiles diverged immediately after switching to PC, where the cell retained the iron anode (supplying Fe^{2+} for Fenton chemistry) but employed either graphite or BDC as the cathode. With the BDC, TOC removal rose rapidly from 30% to 60% within the first 15 min of PC and continued to increase to 85–86% by 90–120 min of electrolysis (Fig. 5.12). In contrast, the graphite cathode increased more slowly from 30% to 55% in the first 15 min and plateaued near 76% after 120 min of PC electrolysis (Fig. 5.12). Interpreting the time segments as apparent first-order regimes, the early PC rate with the BDC was 65% higher than with graphite while the later PC rate was 40% higher consistent with the BDC's superior two-electron ORR selectivity and higher local H_2O_2 yield at pH 3.

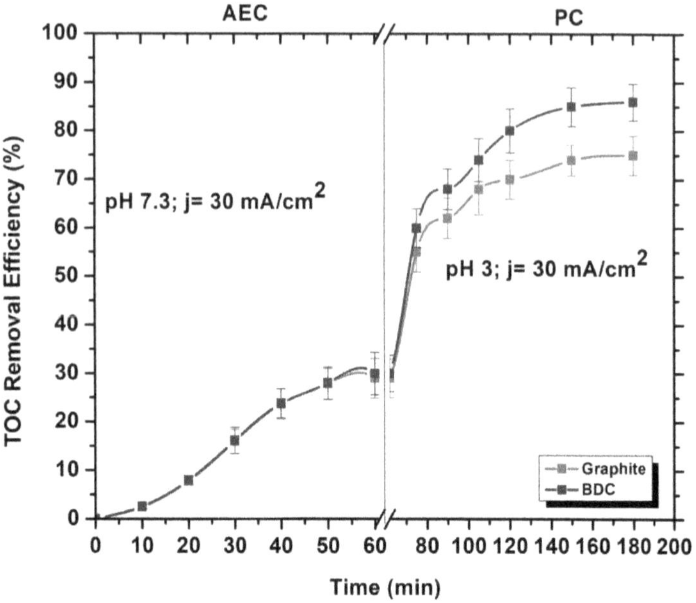

Fig. 5.12 Sequential AEC–PC process for the removal of TOC graphite as cathode and BDC as cathode (AEC: pH 7.3, Current density 30 mA/cm^2, electrolysis time of 60 min. PC: pH 3, Current density 30 mA/cm^2, electrolysis time of 120 min)

In contrast, AEC performance remained comparable because Fe–Fe operation is governed by anodic iron hydrolysis and cathodic hydrogen evolution induced micro-alkalinity which favor $Fe(OH)_3$ sweep-floc formation. These processes are not influenced by the downstream cathode choice. Practically, the BDC configuration achieved 70% TOC removal about 35 minutes earlier than graphite at the same current density, while ultimate removal improved by approximately 10 percentage points without increasing sludge, since the incremental gain resulted from the oxidation step rather than additional coagulation.

The sludge quantity generation in the sequential AEC-PC process with BDC electrode was more or similar to that of the quantity reported for sequential AEC-PC process with graphite as cathode (Section 4.5).

5.6 Summary

In this work the biomass derived cathode electrode has been fabricated from waste teak wood using pyrolysis-based method. The characterisation of the electrode was done by SEM, EDX, Raman analysis, XRD, TGA-DSC, BET, FTIR, EIS and CV. SEM results showed a microchannel porous structure formation in the material, the thermal stability of the material was analysed using TGA-DSC and the pyrolysis temperature of 900 ^{0}C was fixed accordingly. EDX analysis indicated a rise in carbon content with respect to the pyrolysis temperature. Raman analysis detailed that disorder increased as a result of increase in pyrolysis temperature. XRD and FTIR analysis results exhibited the extension of graphitization of the material and CV results showed the increase in electroactive surface area (1.138 cm^2) of BDC electrode compared to the graphite electrode (0.87 cm^2). EIS revealed the lower impedance and charge transfer ability of the BDC electrode. Furthermore, the performance of biomass derived electrode was tested for the treatment of landfill leachate, where the removal efficiency of COD, TOC, AN, TN observed was 30.6%, 37.5%, 32.7% and 34.6% respectively at pH 7.3, current density of 30 mA/cm^2, Voltage 3 V and reaction time of 120 min. The removal of TOC was enhanced to 77% when the pH of the leachate was reduced to 3 in the biomass-derived cathode system. These results are comparable to those achieved with graphite electrodes, underscoring the potential of the BDC electrode as a promising and sustainable cathode material.

www.ingramcontent.com/pod-product-compliance
Lightning Source LLC
LaVergne TN
LVHW051923051126
851857LV00021B/2111
* 9 7 9 8 2 5 6 2 9 3 7 1 0 *